The
PYCNOGENOL®
PHENOMENON

The most unique & versatile health supplement

The role of Pycnogenol®
in human health and disease

PETER ROHDEWALD, Ph.D. AND
RICHARD A. PASSWATER, Ph.D.

D0818563

**Basic
Health**
PUBLICATIONS, INC.

The information contained in this book is based upon the research and personal and professional experiences of the authors. It is not intended as a substitute for consulting with your physician or other healthcare provider. Any attempt to diagnose and treat an illness should be done under the direction of a healthcare professional.

The publisher does not advocate the use of any particular healthcare protocol but believes the information in this book should be available to the public. The publisher and authors are not responsible for any adverse effects or consequences resulting from the use of the suggestions, preparations, or procedures discussed in this book. Should the reader have any questions concerning the appropriateness of any procedures or preparation mentioned, the authors and the publisher strongly suggest consulting a professional healthcare advisor.

Basic Health Publications, Inc.
885 Claycraft Road
Columbus, OH 43230
800-334-9969 • www.basichealthpub.com

Originally published by Ponte Press Verlags-GmbH, 2015, Stockumer Straße 148 44892 Bochum, Germany

Pycnogenol® is a registered trademark of Horphag Research.

Library of Congress Cataloging-in-Publication Data is available through the Library of Congress.
Rohdewald, Peter, author. | Passwater, Richard A., author.
  The pycnogenol phenomenon : the most unique & versatile health supplement / Peter Rohdewald, Ph.D., and Richard A. Passwater, Ph.D.
  Columbus, OH : Basic Health Publications, Inc., [2016]
    LCCN 2015037170 | ISBN 9781591204015
    LCSH: Bioflavonoids—Physiological effect. | Antioxidants.
    LCC QP772.B5 R64 2016 | DDC 613.2/86—dc23
    LC record available at http://lccn.loc.gov/2015037170

ISBN-13: 978-1-59120-401-5

Editor: Dr. Stefan Siebrecht, Health & Nutrition Books II

Printed in the United States of America

10   9   8   7   6   5   4   3   2   1

Contents

Chapter Eight | Living Better Longer: Additional Anti-Aging Effects of Pycnogenol®: Memory Enhancement and Longevity 109

Chapter Nine | Allergies, Asthma and COPD 115

Preface

As scientists, our goal is to help people live better longer. The discovery process is only part of our mission. We will not have fulfilled our mission until we have assured our findings reach beyond the clinical setting and reach those who can benefit. This is what we seek to accomplish with this book.

We will discuss many of the health benefits of the nutritional supplement, Pycnogenol® (pronounced pic-noj-en-all). Some of the major health benefits, such as heart and artery health, protect against degeneration and help protect life and wellness. Pycnogenol® boosts the immune system and helps keep blood vessels and capillaries strong, flexible and open. Other Pycnogenol® benefits support a life full of energy and joy, as well as quality of life concerns such as healthy, younger-looking skin, a healthy sex life and normal, pain-free movement.

Our message is the health benefits of Pycnogenol®, many of which have been elucidated under Professor Rohdewald's leadership. Professor Rohdewald has conducted much of the research in his laboratory as well as guided the modern research in other laboratories.

The Pycnogenol® story begins with the recognition that an extract of pine bark was used by Native North Americans to restore health at least as long ago as the 1500s. A commercial extract from the bark of French Maritime Pine trees was developed in France in 1953 and called Pycnogenol®. The name "Pycnogenol®" is based on Greek words depicting to chemists that the large bioflavonoid nutrients are formed when the bark joins together smaller bioflavonoid molecules. By the late 1970s, Pycnogenol® was known for its anti-inflammatory action, improvement in capillary health, and better skin. Since Professor Rohdewald was a leading pharmaceutical researcher, a physician friend asked him to investigate this patented extract.

As Professor Rohdewald recollects, "My interest in Pycnogenol® began in 1982 when a friend, a medical doctor, entered my office at the university and asked if I spoke French. I answered: Yes, a bit, but, why are you asking? He explained to me that he was interested in a dietary supplement from France that reduces discomfort of seasonal allergies. My physician friend asked the manufacturer for clinical studies but found that they were only in French. He asked if I would make a German language summary of the French language studies. I agreed and he gave me a stack of documents. After a quick read of the studies, I thought it would be a fun idea to travel with this package of French journals to France to spend my holidays there while learning about this French health ingredient at the same time."

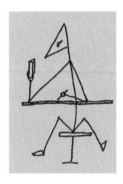

"My family and I traveled to France to the Landes de Gascoigne forest near to the Bay of Biscay region of the Atlantic Ocean. The forest consisted of miles and miles of pine trees. For about two weeks, I was sitting in the shade of pine trees, translating the French documents. I realized during the translation that the thick, brown bark of the pine trees around us, more exactly the French maritime pine trees (Pinus *pinaster atlantica*), was the raw material for Pycnogenol®, which was the subject of all these articles. The nutrients extracted from the bark are a specific blend of active bioflavonoids, also found in fresh fruits and vegetables."

"The forest had been planted by order of Napoleon III to halt the erosion of the sandy heath land by the strong Atlantic winds, to cleanse the soil and to start a timber industry. The pine trees cover now an area of about 3,900 square miles. They flank the beautiful sandy beach from near the vineyards of Bordeaux to the north and the Pyrenees Mountains to the south of Biarritz, the silver coast. This forest is the largest forest in Western Europe and it provides the bark as an abundant source of raw material for the production of Pycnogenol®. It is a totally unspoiled environment, because population density is small, except mid-summer, when tourists invade the beaches."

"The bark is not gathered from uncut trees, but is taken from trees freshly harvested for timber. The forest is managed according to the rules of Good Agricultural Practice. New trees have to be planted after harvesting the large pines, which are used very intensively. The wood of the trees ends up mainly in furniture and construction material. The pine bark is

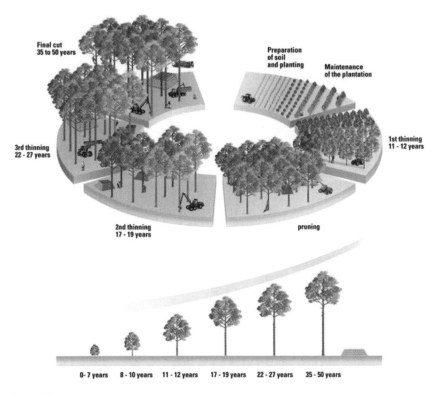

Final cut
35 to 50 years

Preparation
of soil
and planting

Maintenance
of the plantation

3rd thinning
22 - 27 years

1st thinning
11 - 12 years

2nd thinning
17 - 19 years

pruning

0- 7 years   8 - 10 years   11 - 12 years   17 - 19 years   22 - 27 years   35 - 50 years

Forest Cyle

used for the extraction of Pycnogenol®. The extraction plant for the bark is situated in the middle of pristine French forest."

"By the end of our vacation, I had finished the translations and was able to submit the first article about Pycnogenol® to a German medical paper. By now, I was extremely excited and intrigued by Pycnogenol®. During these weeks I had learned that Pycnogenol® had positive effects on eye health, micro-bleedings (pinpoint drops of blood that leaked from blood vessels) and edema (fluid in tissue) of the lower legs. At this time, the clinical studies were mostly detailed case reports. Knowledge about the ingredients of the maritime pine bark and the mode of action was sparse at that time. As a pharmaceutical researcher, this is what interests me most. Thus, I set out to investigate this pine bark extract more deeply, which is what I did during the next ten years."

"During this time we worked in my laboratories at the University of Münster, Germany, on the analysis and bioavailability of Pycnogenol® and were able to elaborate a method to standardize the composition of the complex extract. As a result, the Pharmacopoeia of the United States (USP) describes Pycnogenol® in full detail in the monograph "Maritime pine extract" Since then there have been a great number of studies – more than 100 published clinical studies and 300 scientific publications. The studies started with *in vitro* tests that led to controlled clinical trials. They are the scientific basis that demonstrate the vast potential of Pycnogenol® for human health."

"As my studies continued, I became more and more interested in the health benefits of Pycnogenol®. Eventually, I became a consultant to the manufacturer, Horphag Research, Ltd and upon my retirement from teaching, I helped direct their research program."

Dr. Passwater, a leading nutritional scientist who has never had an affiliation with Horphag Research, Ltd., recalls his first meeting with Professor Rohdewald in 1993. "I had been researching the synergism of antioxidant nutrients since 1959 and when I read a report indicating that Pycnogenol® was a more powerful antioxidant than both vitamin C and vitamin E, I became very interested. After searching the scientific literature for all published studies on Pycnogenol®, I decided to visit the main researchers involved with Pycnogenol® studies. The number of scientific publications was small then, now they number in the hundreds.

"I started with Professor Rohdewald in his laboratory at the University of Münster in Germany during the 1000 year anniversary of the founding of Münster. I also visited with Professor Antti Arstila of the University of Jyvaskyla in Finland during October, 1993 and Professor Miklos Gabor of the Albert Szent-Györgyi Medical University of Szeged, Hungary while he attended a scientific congress in Paris. Professor Arstila was researching Pycnogenol® and skin health, while Professor Gabor was conducting clinical research on Pycnogenol® and capillary health."

"Professor Rohdewald and his grad students reviewed their research with me and walked me through their present studies. Their earlier efforts had been directed towards determining the nutrients in the patented extract and their exact concentrations. Of course, since Professor Rohdewald was a pharmaceutical researcher, they had also investigated the parameters such as absorption, bioavailability, and circulation in the blood, target organs and toxicology. Since several of the bioflavonoid nutrients had not yet been widely studied, they were also investigating the biochemistry of these nutrients.

Professor Rohdewald had already established the safety and efficacy of Pycnogenol® and further whetted my appetite to learn of its powerful antioxidant and anti-inflammatory properties. I then set out to visit the other Pycnogenol® researchers. As it turned out, through the years, I have had the honor and privilege of sharing the podium at several scientific symposia."

This collaboration was decided upon during Dr. Passwater's most recent effort to update the Pycnogenol® story. Although Dr. Passwater had written six books on Pycnogenol®, rapidly expanding research was uncovering important additional health benefits that need to be brought to the public's attention. These newly verified health benefits have become so extensive and rapidly uncovered that a new book is required to update health professionals and consumers alike. Since so much of the modern research has been led by Professor Rohdewald, it seems only logical and fitting that the two co-author this book. This is a common practice when scientists publish their research or edit books.

Professor Rohdewald will describe the research and Dr. Passwater will discuss how it impacts health. We hope the information will help you live better longer.

Professor Dr. Peter Rohdewald, Ph.D.
Münster August 2014

Richard A. Passwater, Ph.D.
Berlin, Maryland August 2014

Chapter One | Pycnogenol's®
Important Health Benefits for You

Numerous clinical studies prove the natural nutritional supplement Pycnogenol® can significantly improve your health. Everyone's health! No matter how poor or how good your current health – the evidence from hundreds of studies over several decades is clear that Pycnogenol® will bring you important health benefits. The evidence is that Pycnogenol® helps maintain wellness by reducing the damage and decay that ages your body and leads to many of the non-germ diseases associated with aging and decreased longevity. Pycnogenol® also can help strengthen your immune system to protect you from germ diseases. Other Pycnogenol® benefits support an active life-style, maintain good vitality, physical strength, mental health and general well-being. This includes the "quality of life" concerns such as energy, healthy, younger-looking skin, a healthy sex life and normal, pain-free movement.

Our concern is that these health benefits are not widely known. Millions already benefit from Pycnogenol®, but the vast majority of people on this planet have not even heard of Pycnogenol®. If you are one of these, then please let us tell you about many of these documented health benefits and how they can help you live *better* longer. This is our final goal. As researchers, we have helped elucidate many of these health benefits, but unless they are put into practical use, our efforts will have been in vain. It is of little use to make a discovery if it remains buried in research files.

Surely you must be wondering how one nutritional supplement can have so many health benefits. Well, along the way, we'll discuss in a reader-friendly way, how and why Pycnogenol® has so many important health benefits. As scientists, we are trained to be skeptical and we hope you as readers and consumers are skeptics as well. With close examination, the facts become self-explanatory. We will provide fellow researchers, health professionals and pharmacists with the details they need while doing so in plain language for everyone else.

Figure 1.1. Pine bark

As a form of skepticism, we are often asked how one nutritional supplement can have so many benefits. The short explanation for the multitude of health benefits is that Pycnogenol® is not merely one nutrient, but a precise, complex mixture of scarce nutrients, each of which may have multiple functions. The nutrients of Pycnogenol® are lacking in most diets. They are primarily of a class of phytonutrients (plant derived nutrients) called bioflavonoids. Bioflavonoids are loosely considered to be a large family of thousands of plant compounds, many of which are plant pigments. Bioflavonoids (specifically flavonoids such as the catechins) are "the most common group of polyphenolic compounds in the human diet and are found ubiquitously in plants. It is only recently that scientists have found that bioflavonoids are so vital, yet lacking in modern diets. Bioflavonoids have been generally recognized as having biological significance, but are not yet established as essential nutrients. That may be about to change if scientists in the field have their way.

A leading proponent of establishing essentiality and recommended intakes for several phytochemicals including bioflavonoids is Tuft University's Professor Jeffrey Blumberg. Professor Blumberg is the Director of the Antioxidants Research Laboratory at the Jean Mayer USDA Human Nutrition Research Center on Aging, Professor, Friedman School of Nutrition Science and Policy at Tufts University and teaches pharmacology and nutrition in the Sackler School of Graduate Biomedical Sciences. He points out that, "High dietary intakes of phytochemicals are associated with better maintenance of physiologic function and a lower prevalence of many degenerative conditions in older adults. Understanding how polyphenols like the bioflavonoids reduce oxidative stress and inflammation to impact the pathogenesis of chronic disease presents opportunities for health promotion and alternative therapeutic modalities for an aging population." (Ref. 1)

In the Foreword to "An Evidence-Based Approach to Dietary Phytochemicals," Dr. Blumberg states, "It is worth noting that recent research indicates that some bioflavonoids have the potential to influence the expression of genes, suggesting they can influence fundamental aspects of our cellular function despite their not being "essential" to us... They serve to support our physical well-being as well as our mental state." (Ref. 2)

Professor Blumberg has been advocating that recommended intake levels be established for several of the bioflavonoids since before 2010. Progress has been slow, but it is ongoing. "We need to develop a quantifiable reference for these bioflavonoids. We have an RDA for vitamin C and we can put that on a label and it has meaning. But how are we going to do this for the thousands and thousands of phytonutrients that don't fit into the nutritional deficiency paradigm?" (Ref. 3)

Now that we have shown that bioflavonoids such are found in Pycnogenol® not only provide nourishment, they also affect our genes. They can "turn on" and "turn off" specific genes in such a way as to determine our health. This fact makes it easier for health professionals to understand how Pycnogenol® has such wide and positive effects. The new field of nutrigenomics has taught us that the genes that we are born with do not control our destiny. It is the other way around – *we control our genes* by our diet and lifestyle. We can have bad genes that are never activated because we choose a good diet and lifestyle. Or we can have good genes that are never activated because of a poor diet and /or lifestyle.

These findings are of interest to scientists and health professionals, but consumers are usually more interested in results than theory. To illustrate one of the nutritional health benefits, let's discuss some of the ways in which Pycnogenol® affects heart and artery health in the next chapter. Then, for those interested in how Pycnogenol® works to bring these

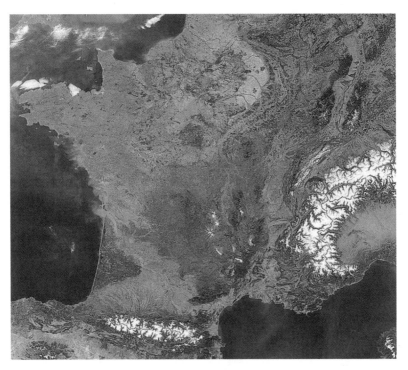

The Landes de Gascoigne forest near to the Bay of Biscay region of the Atlantic Ocean

benefits about, we will – in reader-friendly language – discuss some of its biochemistry in Chapter 3. Of course, you may skip that chapter if you wish. The exciting news is that we have recently uncovered a surprising way in which Pycnogenol® works. But first, let's look at how Pycnogenol® supports heart and artery health by keeping blood vessels and capillaries strong, flexible and open.

Chapter References

1. http://sackler.tufts.edu/Faculty-and-Research/Faculty-Profiles/Jeffrey-Blumberg-Profile on 4/21/2014.

2. J. Higdon, Ph.D., "An Evidence-Based Approach to Dietary Phytochemicals" Thieme Books 2007.

3. http://www.nutraingredients.com/Research/Prof-Blumberg-We-need-reference-intakes-for-phytochemicals-and-we're-not-calling-them-antioxidants on 4/21/2014.

Chapter Two | Pycnogenol®
and Heart Health

Cardiovascular health and vascular function represent the core health benefit of Pycnogenol®. After more than 40 years of research, there is no doubt that Pycnogenol® significantly contributes to a healthier circulation, blood pressure and cholesterol levels. Initially, much of the research was focused on the large arteries and veins. More recently, the focus has been on elucidating in more detail exactly how Pycnogenol® supports the entire cardiovascular system. Pycnogenol® extends its vascular benefits from large blood vessels also to the smaller blood capillaries. A great many ailments are related to insufficient blood microcirculation through the tiniest blood vessels that deliver the oxygen and nutrients to every part of the body to maintain healthy energy levels, normal metabolism, mental concentration and well-being.

Heart health involves not only the heart and the coronary arteries that supply blood directly to the heart, but the entire cardiovascular system of the heart and all arteries. Poor heart health can lead to several types of cardiovascular disease of which coronary heart disease (CHD) is the most common. In CHD, the myocardium (heart muscle) no longer receives sufficient oxygen for its needs. When there is insufficient oxygen for heart needs, it is called myocardial ischemia or ischemic heart disease (IHD). If the lack of oxygen is severe, the myocardium can't produce energy and the oxygen-starved heart muscle cells die, causing a heart attack, which health professionals call a myocardial infarction (MI).

The primary cause of CHD is atherosclerosis, which is the result of atheromatous plaque. The plaques are often called "cholesterol deposits" though they consist of lipids, proteins, cell debris and cholesterol.

Pycnogenol® significantly contributes to the improvement of cardiovascular risk factors due to normalization of blood platelet function and blood-pressure, reducing inflammation in the arteries, maintain the

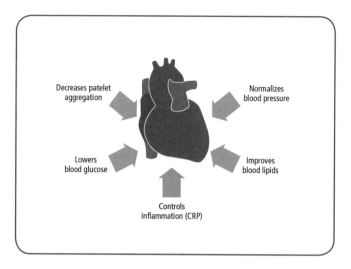

Figure 2.1. Pycnogenol® improves major risk factors for cardiovascular health.

health of artery linings (endothelial function), improvement of blood fats and cholesterol-related components, as well as blood sugar values. Modern research has convinced many health professionals that endothelial function and platelet function are more important factors in heart health than traditional risk factors such as blood cholesterol levels. Reducing the level of chronic inflammation in the body reduces the risk of developing plaques in arteries. A review of the science linking chronic inflammation to atherosclerosis is presented by Dr. J. Keaney Jr. of the University of Massachusetts Medical School. (Ref. 1)

The good news is that Pycnogenol® positively impacts all of these factors. Please see Figure 2.1.

A steadily increasing number of clinical studies demonstrates the efficacy of Pycnogenol® for keeping cardiovascular health problems at bay. Pycnogenol®'s role in helping to keep key parameters in normal ranges has been studied in healthy people, individuals with borderline high risk factors but also as an adjunct in people taking prescription medicine for cardiovascular health issues. Rather than review these studies chronologically, we will group them into categories according to their impact on heart health; platelet function, endothelial function; blood cholesterol, blood pressure and diabetes.

Blood Clots Cause Heart Attacks

Atherosclerosis and hypertension damage the heart slowly, but thrombus (blood clot) formation is an immediate risk. A heart attack rarely occurs in the absence of having a blood clot form in a coronary artery. A spasm of the coronary artery that shuts off blood flow or causes the heart to fibrillate and be unable to pump blood can also result in a heart attack, but the vast majority of heart attacks are caused by a blood clot that forms in a coronary artery that has been constricted by plaque buildup. The majority of strokes are also caused by a blood clot—in this case, a blood clot that forms in a blood vessel that supplies the brain.

The path to a heart attack is a two-step process. Atherosclerosis does not by itself cause heart attacks, but can result in heart pains (angina) due to restriction of the blood flow. (Angina is usually provoked by some sort of physical exertion such as shoveling snow, which increases the heart muscle's demand for oxygen.) Blood clots and vasoconstriction (constriction and/or spasm of an artery) are the events that usually precipitate a heart attack. The narrowing of an artery begins with an inflammatory response where foam cells of the immune system build up under the artery lining. (Foam cells are white blood cells filled with oxidized LDL. LDL are cholesterol carriers in the blood and are discussed later.) These cells promote the infiltration of various substances through the artery wall into its middle layer.

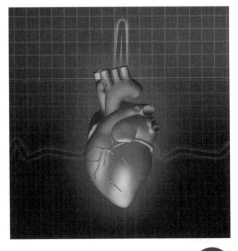

Arteries are said to be "diseased" or "dysfunctional" as plaque is formed in their walls. As the plaque continues to increase, the wall is pushed into the lumen (the opening where the blood flows through) thus narrowing it. This decreases blood flow to the heart tissue. The narrowed artery also affects the blood platelets squeezing by, making the blood sticky and encouraging clot formation at the plaque site. We'll discuss artery endothelial function shortly, but first let's examine the role of blood platelets in clot formation.

Blood Platelets and Clots

Platelets play a major role in the process of blood coagulation to arrest bleeding. When bleeding begins, the vessel constricts, and a protein called *tissue factor* is released by the blood vessel wall. Then *platelet factor 3* is activated on the surface of the platelet that reacts subsequently with blood factors to promote the formation of a platelet plug and initiates other steps in the blood-clotting mechanism. When platelet factor 3 is activated, the platelet changes shape and is said to be "activated."

Platelets, however, can be activated even when there is no bleeding, and this is not good. If they are activated, they still tend to aggregate or clump together and initiate an undesirable blood clot, which can block blood flow through the vessel and result in a heart attack or stroke. If this undesirable blood clot is stationary, it is called a thrombus, and if it travels through the vessel, it is called an embolism. An embolism can reach a narrower vessel and become a thrombus. Platelets can also be activated by smoking, stress, diabetes, and certain nutritional deficiencies. In addition, as people age, a greater percentage of their platelets tend to be undesirably activated.

A critical factor in preventing a heart attack then is to maintain the proper slipperiness of blood cells by maintaining the blood platelets within their normal range of activation. This prevents deadly blood clots from forming in the coronary arteries.

One of the first lines of research with Pycnogenol® and heart health involved its role in maintaining healthy blood platelet function. Professor Rohdewald knew from studies done in Taiwan that procyanidins, a family of nutrients found in Pycnogenol®, inhibit platelet aggregation by blocking the synthesis of thromboxane. (Ref. 2) Thromboxane promotes clot formation.

Professor Rohdewald reasoned that Pycnogenol®, which includes a concentrate of procyanidins, should act in the same way to protect against platelet aggregation. He designed and directed studies in Germany and the USA to see if that was indeed the case. It was. As Professor Rohdewald commented later, "It was very satisfying from a scientific point of view that in our studies, the Taiwanese results, obtained in the test tube, could be transferred successfully to human beings."

We'll discuss these studies shortly, but first, let's consider how smoking – another risk factor in heart disease – does its damage.

Thrombosis: One Risk of Smoking

Professor Rohdewald reasoned that besides cardiovascular diseases and diabetes there is another great opportunity to promote undesired blood clotting. Smokers are at a higher risk for heart attacks and strokes. The inhaled and absorbed smoke from just one cigarette stimulates the aggregation of blood platelets.

Professor Rohdewald also realized that the cigarette-induced activation of blood platelets, leading to aggregation of blood platelets, was an outstanding test system. He noted, "Volunteers provoke the aggregation of blood platelets with pleasure, because they like to smoke, and results could be obtained within a short period of time."

Professor Rohdewald directed studies at universities in the USA as well as in Germany. In 1997 he interested Professor Ronald Watson of University of Arizona, a leading antioxidant researcher, in participating in some of the studies. The results were impressive. The smoking of 3 cigarettes within 30 minutes activated platelets, so that 20% more platelets were aggregated. This aggregation could be completely prevented by Pycnogenol®, depending on the dosage. It is a long lasting effect; even 6 days after intake of Pycnogenol® the blood platelets were less aggregated than before. (Ref. 3)

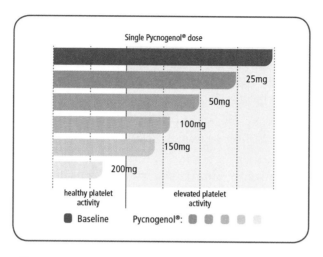

Figure 2.2. Pycnogenol® inhibits platelet aggregation after smoking.

Pycnogenol® and Bleeding

It is certainly desirable to prevent unwanted platelet activation as the first step of unwanted clot formation. However, there is the normal, life-saving function of blood platelets that is to stop bleeding and this function should not be impeded or inactivated. To control whether Pycnogenol® interferes with this important platelet function, the bleeding time was noted in parallel to the platelet aggregation. No increase of bleeding time was noted, whereas aspirin, which also prevents platelet activation, prolonged bleeding time considerably. (Ref. 3)

As Professors Rohdewald and Watson noted, "The comparison with aspirin in our study revealed that Pycnogenol® has the same anti-aggregatory effect, but obtained with a lower dose, and, most important, no unwanted increase of bleeding time. Thus Pycnogenol® reduces the tendency of blood to form unwanted blood clots without the risk of gastric bleedings, which is a common side effect of aspirin."

"There are several ways how Pycnogenol® blocks platelet activation. The investigators from Taiwan found an inhibition of thromboxane formation by procyanidins. (Ref. 2) Pycnogenol®, with procyanidins as main ingredients, should block the formation of thromboxane, too. Smoking increased thromboxane levels in our studies. However, intake of Pycnogenol® was able to bring down thromboxane levels to normal, similar to nonsmokers."(Ref. 4)

The ability of Pycnogenol® to help keep blood platelet activation within the normal range reduces a major – if not the number one – risk of heart attack. However, there is an earlier condition of the arteries that must be maintained – a healthy endothelial function. A chronic "silent" background inflammation can affect endothelial function and lead to the formation of plaque in the arteries.

Inflammation and Heart Health

Chronic inflammation is now recognized by many experts for heart diseases as being the major underlying cause of heart disease. This shift in medical thinking away from blood cholesterol as being the most important factor is illustrated by a seminal report in the American Heart Association (AHA) journal *Circulation* in 2002. (Ref. 5)

To quote from the article's abstract, "Atherosclerosis, formerly considered a bland lipid storage disease, actually involves an ongoing inflammatory response. Recent advances in basic science have established a fundamental role for inflammation in mediating all stages of this disease from initiation through progression and, ultimately, the thrombotic complications of atherosclerosis. These new findings provide important links between risk factors and the mechanisms of atherogenesis [the process of plaque formation in the inner lining of the arteries]. Clinical studies have shown that this emerging biology of inflammation in atherosclerosis applies directly to human patients."

Pycnogenol® Helps Normalize Inflammation

Many chronic inflammatory conditions cause much damage throughout the body. Immune cells produce many reactive compounds called free radicals that ultimately destroy many of the body's own cells. Immune cells perceive the situation as if it were an infection. This calls for more immune cells, which at the same time get more aggressive. This vicious circle explains why many chronic inflammatory disorders persist for so long. Pycnogenol® can help to interrupt the sinister process; it is anti-inflammatory.

Pycnogenol® acts on the master-switch NF-kappa B, which triggers inflammation.

However, this inflammatory switch NF-kappa B is not all evil; we depend on it for every kind of infection. Shutting off this switch completely would cause immune-suppression, something we know from people with organ transplants who take such drugs to prevent tissue rejection. In studies carried out with students who took Pycnogenol®, the activity of NF-kappa B in their immune cells was decreased by 15% (Ref. 6). This may at first thought not sound like an impressive effect, but it does indeed efficiently moderate immune responses. With Pycnogenol®, immune cells can better distinguish between a desirable inflammation such as when we catch flu and a superfluous inflammation such as in the case of arthritis.

Pycnogenol® and Healthy Arteries

Fundamentally, it is the innermost cell layer in blood vessels, the endothelium, which is a major key to heart health. The endothelial

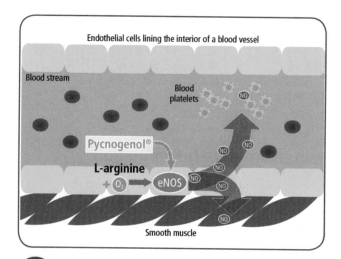

Figure 2.3. Pycnogenol® improves endothelial function.

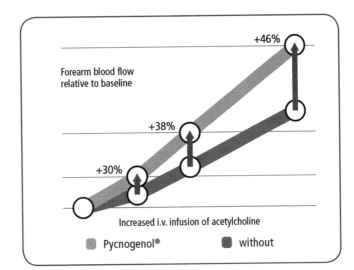

Figure 2.4
Pycnogenol®
stimulates
endothelial-
mediated
vasorelaxation.

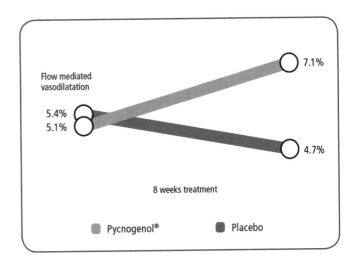

Figure 2.5.
Pycnogenol®
improves
widening of
blood vessels.

cells cover every blood vessel in every organ and they act in concert to regulate blood flow. Endothelial function is now recognized as an important factor in cardiovascular health. Endothelial function can be affected by diet.

The blood flow is regulated by secretion or synthesis of substances that dilate or constrict blood vessels. The balance between these factors is of pivotal importance for heart health. Oxidative stress, diabetes, high cholesterol and inflammation damage the endothelium, leading to con-striction or blocking of blood vessels, to atherosclerosis, hypertension and thrombosis.

The common denominator of most cardiovascular health parameters is the ability of blood vessels to generate the most important vascular media-tor, nitric oxide (NO). NO is continuously produced in the endothelial cells lining the artery walls. It diffuses through the wall, it interacts with a specific receptor in the muscle cells of the artery causing relaxation and consequently, an increased open artery diameter. This facilitates blood flow and normalizes blood pressure. NO also acts on blood platelets to reduce their activity and diminish their stickiness, thus lowering platelet aggregation.

Clinical trials demonstrate that Pycnogenol® stimulates the secretion of two blood vessel relaxing substances: Nitric oxide (NO) and prostacyclin. (Ref. 7) They both, in turn, inhibit blood vessel constricting substances such as endothelins.

Prostacyclin is an effective vasodilator; it makes blood vessels wider thus lowering blood pressure. Endothelins, in contrast, are proteins that con-strict blood vessels and raise blood pressure. They are normally kept in balance with vasodilators, but when they are over-expressed, they con-tribute to high blood pressure and heart disease.

In the endothelium, Pycnogenol® acts as a catalyst on the enzyme that synthesizes NO. In the presence of Pycnogenol®, the endothelial nitric oxide synthase enzyme more efficiently generates NO from its precur-sor amino acid, arginine. (Ref. 8) The increased generation of NO by Pycnogenol® translates into a broad range of cardiovascular health ben-efits. (Ref. 9)

Basically, the nutritional action of Pycnogenol® in helping to maintain NO synthesis inside the endothelium is as important as its ability to maintain proper blood platelet function. It can be argued which is the more important, but both are indeed important. This was proved with young healthy Japanese volunteers. (Ref. 10)

After two weeks with Pycnogenol®, the endothelium produced a 46% increase of blood flow in the forearm. When production of NO was blocked, Pycnogenol® could not dilate arteries. This had been found some years before with arteries of rats. (Ref. 8)

Also in persons with coronary artery disease, Pycnogenol® improves significantly vasodilatation in double-blind study in a university clinic in Switzerland. (Ref. 11)

Furthermore, an increase of microcirculation was seen in Chinese geriatric patients with heart problems. (Ref. 12)

Thus, the nutritional supplement Pycnogenol® improves endothelial function, supports the widening of blood vessels, and helps maintain blood pressure within normal limits.

Pycnogenol® and Cholesterol

High blood cholesterol levels have long been considered a risk factor for cardiovascular health. Cholesterol is both a steroid and a lipid that is vital for cell membrane stabilization, producing sex hormones and vitamin D, among other functions. Cholesterol is absorbed as a part of our meals and transported to the liver. Our cells can also produce cholesterol themselves, but the body finds it more efficient to store and to produce large quantities of cholesterol in the liver and then transport it to other cells.

Since cholesterol is fat-like (a lipid), it is insoluble in blood, which is water-based. The body assembles particles called lipoproteins because they contain both proteins and lipids to transport fats and cholesterol in the blood. Important lipoproteins are classified as the high-density lipoproteins (HDL), and low-density lipoproteins (LDL).

LDL and HDL carry cholesterol in their lipid interiors while their protein shells enable them to be carried in the bloodstream. LDL carries cholesterol to cells from the liver, while HDL clears unused cholesterol back to

the liver. For decades, LDL has been referred to as the "bad cholesterol" and HDL as the "good cholesterol." However, recent research has found that it is not quite that simple as both LDL and HDL particles come in various sizes, composition, and densities that give them different properties. In general, high levels of HDL tend to reduce the risk of cardiac events, whereas high LDL tends to increase the risk of atherosclerosis, stroke and heart attack.

High levels of LDL may lead to atherosclerosis when the cholesterol they carry becomes oxidized to peroxides when LDL antioxidant protectors are inadequate.

Several clinical trials have investigated the effect of Pycnogenol® on blood cholesterol levels.

Taking Pycnogenol® for six weeks significantly increased HDL and lowered LDL in healthy American volunteers. (Ref. 13) Even four weeks after the end of the test period, their HDL was still high, while LDL was back to the values at start.

In a German study with 40 patients showing signs of serious hypercholesterolemia (high blood cholesterol), Pycnogenol® significantly lowered LDL and total cholesterol levels while increasing HDL levels. (Ref. 14)

A Slovakian study with men with mild hypercholesterolemia came to the same results: 3 months after intake of the pine bark extract, both total cholesterol and LDL were significantly decreased, HDL levels were elevated. (Ref. 15)

In a study with 200 women LDL was significantly decreased after 6 months under Pycnogenol®. HDL was significantly increased. (Ref. 16)

LDL was also lowered in the American study with diabetic patients. (Ref. 17)

The relationship between LDL and HDL is used as the atherosclerotic index. As the above listed studies show, Pycnogenol® shifts the atherosclerotic index towards a better prognosis.

The raise of HDL and fall of LDL signalize that Pycnogenol® helps maintain healthy arteries and according to widely accepted theory, help protect against atherosclerosis.

Hypertension and Diabetes

Sometimes, results, obtained on laboratory animals tell us remarkable stories. Very sophisticated experiments with old mice conducted at the University of Arizona demonstrated the great impact of Pycnogenol® on heart health. (Ref. 18)

Old mice had developed heart failure because of high blood pressure (hypertension). Heart failure doesn't mean that the heart isn't working at all, but it is a term to describe that the heart is not pumping sufficient blood for normal health. Physicians often call this condition Congestive Heart Disease (CHD).

The heart failure was associated with changes in the cardiac tissue, driven by the heart's production of large quantities of tissue-destroying enzymes called proteases. The high blood pressure had caused the heart to produce the protease enzymes. The heart tissue (mainly the protein collagen) was attacked by the protease enzymes, thus weakening the heart structure, resulting in insufficient blood being pumped to the rest of the body.

The mice with the heart failure were given Pycnogenol® for one month. Thereafter, impressively, their blood pressure was normalized, protease enzymes in their heart tissues decreased and collagen synthesis increased. Most important, following treatment with Pycnogenol®, all cardiac functions were the same as for control mice without heart failure, in contrast

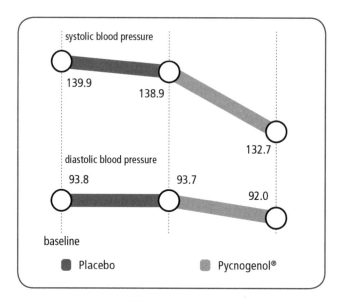

Figure 2.6. Pycnogenol® helps improve blood pressure.

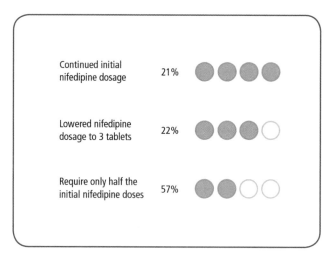

Figure 2.7. Pycnogenol® reduces need for antihypertensive drugs.

to the untreated group with still persisting heart failure. The very remarkable result of the study is that Pycnogenol® was able to restore all functions of the damaged hearts to normal. (Ref. 18)

Normalization of Blood Pressure

Several clinical studies verified that Pycnogenol® is able to reduce the cardiac risk factor hypertension. In a pilot study with American patients with mild hypertension, Pycnogenol® lowered blood pressure after 6 weeks. (Ref. 19)

Monitoring blood pressure in 100 Taiwanese women showed also a lowering of blood pressure after 6 months. (Ref. 16)

In clinical studies from China (Ref. 20) and USA (Ref. 17), patients could reduce the number of anti-hypertensive tablets, nevertheless, blood pressure was lowered.

In two studies in Italy, it could be shown that addition of Pycnogenol® to an antihypertensive treatment could further reduce blood pressure. (Refs. 21 and 22)

These studies with hypertensive patients showed that Pycnogenol® was normalizing mild hypertension and was also helpful as adjunct to standard hypertensive medication. There are several causes of high blood pressure, some of which require medication. Here, too, Pycnogenol® can still be a benefit as it can reduce the amount of medication required.

Pycnogenol® Helps Maintain Heart Health in Several Ways

A USA study successfully demonstrated the drug-sparing effect of Pycnogenol® in persons suffering from the combination of hypertension and diabetes.

With the nutritional support of Pycnogenol®, half of the subjects could reduce the dose of their high blood pressure medication, and blood pressure was nevertheless lowered. The function of the endothelium was improved as indicated by significantly lower levels of the potent blood vessel constrictor, endothelin-1. Because of the lower blood pressure, the kidneys could retain proteins more properly so that urinary albumin was lowered. Last, but not least, blood glucose was better controlled in the Pycnogenol® group. This investigation provided a good example that Pycnogenol® can help maintain normal levels of key factors that reduce important cardiovascular risk factors simultaneously: hypertension and diabetes. In addition, a better kidney function was observed. (Ref. 17)

An Italian study with 58 patients suffering from the metabolic syndrome (Syndrome X) underlined these findings impressively. Metabolic syndrome is defined as having three out of five of the following conditions: abdominal obesity, elevated blood pressure, elevated fasting blood sugar, high blood triglycerides, and low high-density cholesterol (HDL) levels. Metabolic syndrome increases the risk of developing cardiovascular disease, particularly heart failure, and diabetes.

Also in this investigation, the antihypertensive medication could be reduced in the Pycnogenol® group, as well as their kidney function and blood sugar control significantly improved. (Ref. 23) In Chapter Four, we will discuss the many benefits of Pycnogenol® in the nutritional support of diabetics.

One may conclude that Pycnogenol® helps reduce the damage to the heart caused by diabetes and hypertension, two important factors for cardiac risk.

When Professor Rohdewald mentioned during a 1996 lecture in China that Pycnogenol® should be good for heart health, because it makes blood vessels wider, a prominent Chinese medical doctor, became interested.

She decided to test Pycnogenol® on elderly patients to see whether she could find positive effects on the cardiovascular system. She enrolled 60 patients within a short period of time to start a double-blind, placebo-controlled investigation of persons suffering from coronary heart diseases. (Ref. 12)

She found health benefits and no more side effects than that experienced with those taking the placebo. She found that the geriatric patients tolerated Pycnogenol® very well.

One result of the study was to confirm the earlier studies on blood platelet health. Pycnogenol® inhibited the activation of blood platelets. Geriatric patients with cardiovascular diseases have a high risk of having a stroke. On the contrary, the patients belonging to the Pycnogenol® group were to some extent protected against thrombosis during the investigation. In blood from subjects in the Pycnogenol® group, less aggregates of blood platelets were formed in four different test systems, compared to placebo control. So the risk of unwanted blood clots was lowered by Pycnogenol®. (Ref. 12)

How Does Pycnogenol® Do All of This?

We have discussed some of the evidence that Pycnogenol® helps maintain heart health by maintaining healthy blood platelets; controling silent inflammation; helping maintain cholesterol within normal levels; favorably impacting blood pressure; and reduing the damage to the heart from free radicals, smoking, high blood pressure, and diabetes. They are summarized in Table 2.1.

Table 2.1: Four Major Cardioprotective Actions of Pycnogenol®

	1	2	3	4
	Anti-Clotting	**Anti-inflammatory/Cholesterol lowering**	**Endothelial protection**	**Blood sugar regulation**
What Pycnogenol® does to reduce risk.	Normalizes blood platelet activation.	Reduces chronic inflammation and reduces bad cholesterol increases good cholesterol.	Stimulates production of nitric oxide in artery walls.	Helps maintain proper blood sugar levels and reduces oxidative damage.
How	Blocks the production of thromboxane.	Helps regulate the inflammation master switch NF-kB.	Acts as catalyst for the enzyme that makes NO.	Improves insulin sensitivity and reduces free radicals.
Result	Slippery blood, free of blood clots.	Less inflammation, less risk of atherosclerosis.	More open and flexible blood vessels.	Less damage to heart, eyes and nerves.

Hopefully by now, you are becoming interested in how the vital nutrients in Pycnogenol® that few people have heard of can bring all of these health benefits. If so, Chapter Three will discuss the basic biochemistry details that will explain this. If you still aren't interested in the biochemistry, then skip the next chapter and go on to read about more of the many health benefits that are important to you.

Chapter References

1. Keany IF. Immune modulation in atherosclerosis circulation. Circulation 124:559–560, 2011

2. Chang WC, Hou FL. Inhibition of platelet aggregation and arachidonate metabolism in platelets by procyanidins. Prost Leuk Essen Fatty Acids 38: 181–188, 1989

3. Pütter M, Grotemeyer KHM, Würthwein G, et al.   Inhibition of smoking-induced platelet aggregation by aspirin and Pycnogenol®. Thromb Res 95 7: 14–18, 1999

4. Araghi-Nicknam M, Hosseini S, Larson D, et al. Pine bark extract reduces platelet aggregation. Int Med 2: 73–77, 1999

5. Libby P, Ridker PM, Maseri A. Inflammation and atherosclerosis. Circulation 105: 1135–1143, 2002

6. Grimm T, Chovanova Z, Muchova J, et al. Inhibition of NF-κB activation and MMP-9 secretion by plasma of human volunteers after ingestion of maritime pine bark extract (Pycnogenol®). J Inflammation 3:1, 2006

7. Liu X, Wei J, Tan F, et al. Pycnogenol® French maritime pine bark extract, improves endothelial function of hypertensive patients.Life Sci 74: 855–862, 2004

8. Fitzpatrick DF, Bing B, Rohdewald P. Endothelium-dependent vascular effects of Pycnogenol®. J Cardiovasc Pharmacol 32: 509–515, 1998.

9. Watson RR. Pycnogenol® and cardiovascular health. Evid Based Integrative Med 1: 27–32, 2003

10. Nishioka K, Hidaka T, Nakamura S, et al. Pycnogenol®. French Maritime Pine Bark Extract, augments endothelium-dependent vasodilation in humans. Hypertens Res 30: 775–780, 2007

11. Enseleit F, Sudano I, Périat D, et al. Effects of Pycnogenol® on endothelial function in patients with stable coronary artery disease: a double-blind,

randomized, placebo-controlled, cross-over study. Eur Heart J 33: 1589–1597, 2012

12. Wang S, Tan D, Zhao Y, et al. The effect of Pycnogenol® on the microcirculation, platelet function and ischemic myocardium in patients with coronary artery diseases. Eur Bull Drug Res 7: 19–25, 1999

13. Devaraj S, Vega-Lopéz S, Kaul N, et al. Supplementation with a pine bark extract rich in polyphenols increases plasma antioxidant capacity and alters plasma lipoprotein profile. Lipids 37: 931–934, 2002

14. Koch R. Comparative study of Venostasin® and Pycnogenol® in chronic venous insufficiency. Phytother Res 16: 1–5, 2002

15. Durackova Z, Trebaticky B, Novotny V, et al. Lipid metabolism and erectile function improvement by Pycnogenol®, extract from the bark of Pinus pinaster in patients suffering from erectile dysfunction – a pilot study. Nutr Res 23: 1189–1198, 2003

16. Yang HM, Liao MF, Zhu SY, et al. A randomized, double-blind, placebo-controlled trial on the effect of Pycnogenol® on the climacteric syndrome in perimenopausal women.Acta Obstet Gynecol Scand 86: 978–985, 2007

17. Zibadi S, Rohdewald P, Park D, et al. Reduction of cardiovascular risk factors in subjects with Type 2 Diabetes by Pycnogenol® supplementation. Nutr Res 28: 315–320, 2008

18. Zibadi S, Yu Q, Rohdewald P, et al. Impact of Pycnogenol® on cardiac extracellular matrix remodeling induced by L-NAME administration to old mice. Cardiovasc Toxicol 7: 10–18, 2007

19. Hosseini S, Lee J, Sepulveda RT, et al. A randomized, double-blind, placebo-controlled, prospective, 16 week crossover study to determine the role of Pycnogenol® in modifying blood pressure in mildly hypertensive patients. Nutr Res 21: 1251–1260, 2001

20. Liu X, Wei J, Tan F, et al. Pycnogenol® French maritime pine bark extract, improves endothelial function of hypertensive patients. Life Sci 74: 855–862, 2004

21. Cesarone MR, Belcaro G, Stuard S, et al. Kidney Flow and Function in Hypertension: Protective Effects of Pycnogenol® in Hypertensive Participants – A Controlled Study. J Cardiovasc Pharmacol Ther 15: 41–46, 2010

22. Stuard S, Belcaro G, Cesarone MR, et al. Kidney function in metabolic syndrome may be improved with Pycnogenol®. Panminerva Med 52 (suppl. 1 to No. 2): 27–32, 2010

24. Belcaro G, Cornelli U, Luzzi R, et al. Pycnogenol® supplementation improves health risk factors in subjects with metabolic syndrome. Phytother Res 27: 1572–1578, 2013

Chapter Three | Biochemistry of Pycnogenol®

Pycnogenol® is a unique complex plant extract that provides many health benefits. Many of its vital nutrients are in very short supply in modern diets. It is helpful to know what nutrients are present in Pycnogenol® and how they work in the body. To help understand this, let's briefly discuss the nutrient composition of Pycnogenol®. If you are not interested in this, then simply skip this chapter and jump to Chapter Four where we examine Pycnogenol® and bone health. You can always come back to this chapter if and when you wish.

In Chapter One, we pointed out that Pycnogenol® is a precise, complex mixture of scarce nutrients primarily of a class of phytonutrients called bioflavonoids. Bioflavonoids are loosely considered to be a large family of thousands of plant compounds, many of which are plant pigments. Bioflavonoids (specifically flavonoids such as the catechins) are the most common group of polyphenolic compounds in the human diet.

We quoted Professor Jeffrey Blumberg as stressing, "Understanding how polyphenols like the bioflavonoids reduce oxidative stress and inflammation to impact the pathogenesis of chronic disease presents opportunities for health promotion and alternative therapeutic modalities for an aging population. (Ref. 1) OK, now let's examine these mechanisms as they apply to Pycnogenol®.

Professor Rohdewald had already identified and characterized about 85% of the constituents in Pycnogenol® by the time of Dr. Passwater's 1993 visit to his laboratory. The monograph "Maritime Pine Extract" of the US Pharmacopoea describes the composition of Pycnogenol® in detail, as Pycnogenol® corresponds to the standards of the US Pharmacopoea. (Ref. 2) All of the substances contained in Pycnogenol® are found also in other plants that have been used in human nutrition over centuries. The main constituents are natural antioxidants called procyanidins, which are also contained in foods such as sorghum, avocado, strawberries, bananas, and others. The procyanidins themselves are a family of compounds dif-

fering in structure and chain length, but all composed of catechin or epicatechin units.

Procyanidins Are Unique Nutrients

Pycnogenol®'s nutrients come in a range of molecular sizes. The medium-to-large size molecules of Pycnogenol®, the procyanidins, are the most important. Procyanidins have absolutely nothing in common with the poison cyanide. They got their name from a natural colorful pigment, cyanidin, which is the basic constituent of the beautiful color of the cornflower (*Centaurea cyanus*). The procyanidins produce a deep red color when they are heated with acids. Many well-known fruits and berries, as for example apples, grapes, blackberries, contain procyanidins. They have been part of our diet and daily nutrition from the beginning of human existence.

The procyanidins form biopolymers by joining 2–12 or more molecules of smaller molecules (monomers) called catechin and epicatechin together. Molecules of this size are generally not considered to be polymers, but are called oligomers (from oligo meaning 'few'). These oligomeric procyanidins (OPC) are the main constituents of Pycnogenol®.

Pycnogenol® contains other monomeric nutrients such as taxifolin and phenolic acids in addition to catechin and epicatechin. These various monomeric nutrients are also part of our normal daily nutrition. Apples, plums, cherries, apricots contain, for example, p-cumaric acid, caffeic acid and ferulic acid, which are constituents of Pycnogenol®, but in minor quantities. Catechins are familiar to our body as they are also contained in green or black tea. Another component of Pycnogenol® is the flavonoid taxifolin. It is found in grapefruit and oranges. The various monomeric nutrients are antioxidants and anti-inflammatory agents and possess beneficial properties for the human blood circulation. They are present in Pycnogenol® in low quantities.

Bioavailability

Pycnogenol® is one of the most studied and best documented plant extracts in the world and over the last decades many details were discovered about the metabolic fate of Pycnogenol® in the human body. Detailed information about the mechanism of Pycnogenol® in our organism had been elaborated in the laboratories of the University in Würzburg, Germany.

Following the ingestion of Pycnogenol®, all the constituents pass the stomach intact and reach the gut. The small molecules are absorbed in the upper intestine and are released into the blood stream.

They are partly bound to the proteins of the blood plasma and to a certain extent are modified by esterification (reacting acids with the alcohol groups). Within the first six hours the monomers and phenolic acids are excreted in urine. (Ref. 3)

The larger molecules in Pycnogenol® are too big to be absorbed intact, but our body has the ability to metabolize them into smaller entities such as the monomers.

The monomers, mostly epicatechin, catechin and taxifolin, can be absorbed then by the human body or they are further metabolized by the gut microflora into many different metabolites. Here we have our bodies symbiotically working with the nutrients in Pycnogenol® to make even more beneficial substances.

An important discovery has been that to identify one specific metabolite that turned out to be the major metabolite (changed into smaller compounds by the body) of Pycnogenol®. This has been the result of studies in the laboratories of Professor Högger from the University of Würzburg,

Germany. Let's take a closer look at this specific molecule in order to understand why and how Pycnogenol® really works.

The "Active Nutrient" M1

Even though we knew that Pycnogenol® had many health benefits for decades, it was not clear which compounds of the complex maritime pine bark extract are mainly responsible for the documented biological effects until these recent findings.

The major metabolite of Pycnogenol® is called M1, a valerolactone, and it is the most interesting metabolite that is created from Pycnogenol®. (Ref. 4) The presence of more minor metabolites could also be detected, but the structure and properties of these substances are still little known.

M1 inhibits the proteases, which destroy tissues during inflammation, more effectively than hydrocortisone and Pycnogenol® itself. (Ref. 4) That means M1 is a very active metabolite. Furthermore, M1 is a better scavenger of harmful free radicals than vitamin C and Trolox, the vitamin E derivative. (Ref. 4) The pronounced antioxidant activity as well as inhibitory effects upon various matrix proteinases (protein destroying enzymes) are consistent with the reported anti-inflammatory effects of Pycnogenol®. Finally, after all these studies, we can say, that M1 is an active form of Pycnogenol®. M1 is not a primary constituent of pine bark or Pycnogenol®. M1 is generated in the body from the procyanidins' catechin units through multiple step reactions by bacteria inside the lower part of the human bowel. M1 is than released into the blood stream. The digestion and conversion of procyanidins into M1 takes time, therefore, the metabolite M1 appears much later in the blood than the monomeric nutrients. M1 is not detected in the blood until six hours after intake of Pycnogenol® and persists in the blood for up to 24 hours. M1 excretion takes place mainly via urine. (Ref. 5)

In pharmacokinetic studies with single and multiple oral dosages of Pycnogenol®, (Ref. 6) the small molecules catechin, caffeic acid, ferulic acid, and taxifolin were found in the blood plasma of human volunteers, but only in low concentrations.

After M1 was identified, analytical pharmacokinetic studies also found the metabolite M1 in the plasma samples and in urine samples of humans after Pycnogenol® supplementation. The content on M1 in plasma was high enough to inhibit inflammatory substances.

Thereafter, the scientists from the University of Würzburg analyzed also the content of blood cells. They identified a facilitated transport mechanism that led to a pronounced uptake of M1 into endothelial cells, monocytes/macrophages and also into the red blood cells (erythrocytes). (Ref. 7) What we can say today is that M1 from Pycnogenol® travels inside the red blood cells. Via these cells, the active metabolite can reach all organs in our body, including the brain.

Where Pycnogenol® Works in Your Body

Pycnogenol® does not have only one specific target organ where it works. It travels as metabolite with the blood cells and in plasma to every organ, from the brain down to the ankles. Due to its diverse anti-inflammatory actions and its manifold effects on the blood vessels it exerts beneficial effects in many organs.

How Pycnogenol® Works

The health benefits of Pycnogenol® have been intensively studied and verified over the decades. (Ref. 8) The multiple benefits can be explained by the same mode of action that may influence different organs or functions in the same way. Understanding the modes of action of Pycnogenol®, explains why and how Pycnogenol® works and for what health benefits result.

Pycnogenol® Is Anti-inflammatory

Before we ourselves were aware of the health benefits of Pycnogenol®, it was being widely used in several countries in Europe as a nutritional extract to help normalize inflammation. The anti-inflammatory action of Pycnogenol® was what initially sparked Professor Rohdewald's interest in 1983. Pycnogenol® was popular because it helped relieve the symptoms of seasonal allergies (such as hay fever).

When a hypersensitive person comes into contact with an allergen, the body releases histamine in an attempt to fight off the allergen. This release of histamine triggers the symptoms so common to allergies – *inflammation*, sneezing, runny nose, and itchy eyes. The bioflavonoids of Pycnogenol® inhibits histamine *release* from specific body cells (mast cells). We discuss the anti-allergic potential of Pycnogenol® later in chapter Eight.

In Chapter Two, we briefly discussed the role of chronic inflammation in heart disease. Now let's expand the discussion to include other health conditions.

Chronic Systemic Inflammation

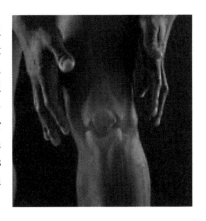

If you fall and scrape your knee, and don't clean the wound thoroughly, it becomes inflamed, red, swollen, warm and painful. The inflammation around your injury is now visible and uncomfortable, so you start taking care of your scrape to minimize the inflammation and promote healing. This incident is considered acute inflammation, which leads to repair of tissue.

When such an injury occurs to the outside of your body, it is easy to identify the source, location and determine treatment. However, when the injury is internal and initially asymptomatic, it may start a chronic inflammation, which is often referred to as *silent systemic inflammation*. We do not see or feel it occurring until it begins to affect our health. Causes of chronic systemic inflammation include: obesity, infections, periodontal disease, environmental toxins, drugs, tobacco, autoimmune diseases and lifestyle choices. If any of the preceding events occur, it frequently results in the onset of vascular damage and disease, hence the nickname 'silent killer' or the 'secret killer.'

Chronic silent systemic inflammation has been now considered as associated with obesity. In obese persons, fat cells grow to gigantic cells. The organism looks at these engorged cells as intruders and tries to eliminate these cells by producing inflammatory substances. This reaction is mostly futile, because our forks and knives continuously create even larger fat cells. So the fight goes on and on with the result that a chronic inflammation takes place. This silent chronic inflammation is not painful, but the inflammatory markers are detectable in the blood. Silent inflammations are a strong underlying cause of many age-related diseases.

Inflammation Is the Root Cause of Many Diseases

The knowledge of the broad anti-inflammatory action of Pycnogenol® helps to understand why it is successful in relieving symptoms of seemingly quite different diseases. Arthritis, endometriosis, asthma, gingivitis, the metabolic syndrome, all these diseases have an inflammatory component. Hence it is no miracle at all that an extract with outstanding anti-inflammatory properties is able to act against so different diseases. The diversity of the diseases is based just on the different location of the inflammation, but they have the same underlying mechanism in common.

Anti-inflammatory Action of Pycnogenol® and Its Metabolites

Pycnogenol® and its metabolites inhibit the inflammation right from the beginning. Experiments with cell cultures showed that Pycnogenol® inhibits the release of many inflammatory markers after adding Pycnogenol® to the cells. However, these cell culture studies do not perfectly reflect reality. Before Pycnogenol® reaches the cells in our body, it has to be absorbed from the intestinal tract. In the gut, components of Pycnogenol® are metabolized (split into pieces and modified), then absorbed after which they travel with the blood stream. Therefore, cells inside our body are exposed to a different mixture of molecules after ingesting Pycnogenol®, compared to the mixture provided by pure (undigested) Pycnogenol® to a cell culture.

Anti-inflammatory Responses After Intake of Pycnogenol®

Following oral consumption of Pycnogenol®, diverse molecules of Pycnogenol® are absorbed by the intestinal tract and released into the blood stream. If human white blood cells were suspended in a serum, one can compare the properties of these cells in blood before and after intake of Pycnogenol®. In such well-designed experiments in Germany it was demonstrated that the blood has the power to inhibit the release of the most important pro-inflammatory triggers after intake of Pycnogenol®. (Ref. 3) First of all, stimulated white blood cells released less NF-κB. The nuclear factor kappa B (NF-κB) commands the liberation of a number of pro-inflammatory molecules; it is the master switch of inflammation.

Inflammatory Components Triggered by NF-κB and inhibited by Pycnogenol®. (Refs. 3 and 4)

Cyclooxygenases (COX) 1 and 2

TNF-α

Interleukins

Prostaglandin-E2

C-reactive protein (CRP)

Matrix metallo proteases (MMPs,) as elastase and collagenase

Pycnogenol® helps to counteract diverse chronic inflammatory diseases by reducing the activity NF-κB. More good news is that there is another positive effect of Pycnogenol® on controlling inflammation. Its antioxidant activity and radical scavenging activity is important for the fight against inflammation, too.

More Than Anti-Inflammatory

Taking Pycnogenol® leads to a remarkable reduction of pro-inflammatory components in the blood. It also greatly reduces excess free radicals in the blood. Both actions are very important to health. While it was the anti-inflammatory action that kindled Professor Rohdewald's interest in Pycnogenol® in 1983, it was the powerful anti-oxidant and anti-radical action that caught Dr. Passwater's attention in 1991. Dr. Passwater was the first scientist to study *antioxidant nutrients* and their synergism in laboratory animals. (Refs. 8-10)

Pycnogenol® scavenges most types of free radicals (reactive oxygen species ROS and reactive nitrogen species RNS) and stimulates the production of anti-oxidative enzymes. A free radical can pull an electron from most other biological compounds, thus restoring its original electron content, but damaging the other compound in the process. It is estimated that each cell in your body (and you have billions of cells) suffers from about 10,000 free radical "hits" each day. The amount of damage depends on how well the cell is protected by antioxidants.

Basic constituents of the body such as DNA, lipids and proteins can be attacked by free radicals and become oxidized. Free radicals are mostly produced inside our body but attack also from outside via environmental factors such as poor nutrition, heavy-metal-containing contaminants, toxins, cigarette smoke, ozone etc. In addition, free radicals may form after our skin is exposed to radiation or UV-light.

Pycnogenol® and Free Radicals

Oxidative stress occurs when, as part of the metabolic process, too many free radical oxygen species are produced within the cell and cannot be neutralized by the body.

The "Free radical theory of aging" postulated by Professor Denham Harman in 1952 suggests that the lifespan of organisms is shortened by an excess of free radicals. Lifespan is determined by several factors, however, it is widely accepted that too much "oxidative stress" is harmful. Free radicals occur in every cell of our body and can attack every part of our body. Free radicals are connected to many non-germ diseases such as we have listed before.

The constituents of Pycnogenol®, especially the procyanidins and its metabolites such as M1, are very potent scavengers of free radicals.

Most of the molecules in Pycnogenol® are able to inactivate more than one free radical per molecule. This is possible because the procyanidins possess more than one reactive group and they are able to retain radical scavenging activity by intra-molecular re-arrangements. (Ref. 11)

Numerous investigations, starting with reactions in test tubes demonstrated the high activity of Pycnogenol® as a scavenger of the most dangerous free radicals.

Pycnogenol®'s Antioxidant Actions Protect Lipids, DNA and Nerve Cells

Radicals attack essential molecules such as DNA and lipids.

Lipids are important because they are needed to form our cell membranes and layers around our nerves to protect them. Lipid oxidation can therefore also damage our nerves.

Common nutritional antioxidants are vitamin C and vitamin E; however, as can be seen in Figure 3.1. Pycnogenol®'s antioxidant capabilities are more powerful than vitamin C and more active than vitamin E or grape seed extract. More on this will be given in Chapter 15: "Pycnogenol® Supplementation."

Studies in Australia and Switzerland with volunteers and patients could both show that Pycnogenol® doesn't only protect lipids against free radicals in the test tube. Pycnogenol® inhibits the oxidation of unsaturated lipids surrounding nerve fibers in men. (Refs. 12, 13)

The inactivation of free radicals may contribute to sustain healthy conditions; more important is the scavenging of free radicals in context with inflammations. The protection of DNA against damage from aggressive free radicals was demonstrated in hyperactive children. (Ref. 14) Pycnogenol® supplementation was able to reduce the damage of DNA in hyperactive children, who expose themselves to oxidative stress by their high level of activity.

Keep the Radicals in Balance

Free radicals are not exclusively noxious for our organism. They serve as signals and they are produced locally to destroy invaders; they are formed during respiration and metabolization of food in large quantities. However, as long as these quantities are in balance with the body's antioxidative factors, there is no harm. But, as in politics, it is the excess of free radicals that is dangerous for the organism. Therefore, taking Pycnogenol® daily, will help to keep the balance.

It should be emphasized that Pycnogenol® not only catches the free radicals, it also stimulates the antioxidative defense systems inside our cells. (Ref. 15)

Our cells possess both an enzymatic antioxidant system and a non-enzymatic antioxidant system. The enzymatic antioxidative system includes catalase (CAT) and superoxide dismutase (SOD). The non-enzymatic anti-oxidative system consists of some vitamins as well as non-vitamin nutrients. The water-soluble vitamins include vitamin C and folic acid. Vitamin E is a fat-soluble antioxidant. Other antioxidant nutrients include coenzyme Q10 and glutathione.

An important finding is that Pycnogenol® more than doubles the antioxidative capacity of cells because Pycnogenol® stimulates the synthesis of several antioxidative agents, such as SOD, catalase and glutathione in cells. (Ref. 15)

Pycnogenol® intake boosts the antioxidative capacity of human blood in several tests. Of highest relevance for our daily life are these results that were obtained when blood of volunteers was analyzed after a period of Pycnogenol® supplementation.

A great variety of tests are available to measure and control the antioxidative capacity of human blood. Using these methods it has been shown in several clinical studies that taking Pycnogenol® increased the antioxidative status of blood of healthy human volunteers.

After six weeks intake of Pycnogenol®, the antioxidant capacity of plasma (ORAC) was reinforced by 40% in 25 healthy overweight Americans. (Ref. 16)

Studies with a free radical absorption system (FRAS) demonstrated a significant decrease of oxidative stress in students (Ref. 17), seniors (Ref. 18), athletes (Ref. 19), and perimenopausal women (Ref. 20) in studies performed at the University of Chiety-Pescara, Italy.

In another study the total antioxidative capacity (TAS) of the blood was elevated in 80 Taiwanese women by 11% following 6 months supplementation with Pycnogenol®. (Ref. 21)

An enhancement of 38% of the biological antioxidant potential (BAP) was found in 78 smokers after intake of 50 mg of Pycnogenol® for eight weeks, relative to placebo. (Ref. 22)

Despite the different methods of measurement, the results point unequivocally to significant improvement of the ability of the human blood to inactivate more free radicals after regular consumption of Pycnogenol®.

Antioxidant Killers: Physical Exertion, Stress, Alcohol, Cigarettes

Sustained high physical exertion (endurance), stress, alcohol and nicotine consumption and many diseases multiply the needs of antioxidants in the body. In smokers, the oxidative stress is dramatic.

Pycnogenol®: An Effective Anti-inflammatory and Powerful Antioxidant

We hope that the foregoing discussion of the basic biochemistry will help you understand the many health benefits of Pycnogenol®. We tried to cover the basics without too much boring biochemistry. The "take home" message is that Pycnogenol® is an effective anti-inflammatory and powerful antioxidant. It has other actions as well as we have shown already in Chapter 2. However, these two properties are themselves synergistic and are responsible for most of Pycnogenol®'s health benefits. Now let's continue our discussion of these health benefits in Chapter Four on Healthy Bones and Joints..

Chapter References

1. Blumberg J, http://sackler.tufts.edu/Faculty-and-Research/Faculty-Profiles / Jeffrey-Blumberg-Profile on 4/21/2014

2. US Pharmacopoea Edition 37 NF 32, Supplements

3. Grosse-Düweler K, Rohdewald P. Urinary metabolites of French maritime pine bark extract in humans. Pharmazie 55: 364–368, 2000

4. Grimm T, Chovanova Z, Muchova J, et al. Inhibition of NF-kappaB activation and MMP-9 secretion by plasma of human volunteers after ingestion of maritime pine bark extract (Pycnogenol®). J Inflamm 3: 1–6, 2006

5. Grimm T, Schäfer A, Högger P. Antioxidant activity and inhibition of matrix metalloproteinases by metabolites of maritime pine bark extract (Pycnogenol®). J Free Radic Biol Med 36: 811–822, 2004

6. Grimm T, Skrabala R, Chovanova Z, et al.Single and multiple dose pharamcokinetics of maritime pine bark extract (Pycnogenol®) after oral administration to healthy volunteers. BMC Clin Pharmacol 6: 4, 2006

7. Kurlbaum M, Mülek M, Högger P. Facilitated Uptake of a Bioactive Metabolite of Maritime Pine Bark Extract (Pycnogenol®) into Human Erythrocytes. PLoS ONE 8:e63197, 2013

8.   Rohdewald P. Pycnogenol®, French Maritime Pine Bark Extract.Encyclo-pedia of Dietary Supplements; Ed. Marcel Dekker, digital publisher, 545–553, 20059. Anon. Chem & Eng. News 48:17 (Oct. 26, 1970)

9.   Passwater, R. A. and Welker, P. Human Aging Research Amer. Lab 3(4) 36–40 (1971)

10. Passwater, R. & Olson, D. US Pat 6, 090,414.

11. Bors W, Michel C, Stettmaier K. Electron paramagnetic resonance studies of radical species of proanthocyanidins and gallate esters. Arch Biochem Biophys 374: 347–355, 2000

12. Enseleit F, Sudano I, Periat D, et al. Effects of Pycnogenol® on endothelial function in patients with stable coronary artery disease: a double-blind, randomized, placebo-controlled, cross-over study. Eur Heart J 33: 1589–1597, 2012

13. Ryan J, Croft K, Wesnes K, Stough C. An examination of the effects of the antioxidant Pycnogenol® on cognitive performance, serum lipid profile, endocrinological and oxidative stress biomarkers in an elderly population. J Psychopharmacol 22: 553–562, 2008

14. Chovanova Z, Muchova J, Sivonova M, et al. Effect of polyphenolic extract, Pycnogenol®, on the level of 8-oxoguanine in children suffering from atten-tion deficit/hyperactivity disorder. Free Radic Res 40: 1003–1010, 2006

15. Nelson AB, Lau BHS, Ide N, Rong Y. Pycnogenol® inhibits macrophage oxidative burst, lipoprotein oxidation and hydroxyl radical-induced DNA damage. Drug Dev Ind Pharm 24: 139–144, 1998

16. Devaraj S, Vega-López S, Kaul N, et al. Supplementation with a pine bark extract rich in polyphenols increases plasma antioxidant capacity and alters plasma lipoprotein profile. Lipids 37: 931–934, 2002

17. Luzzi R, Belcaro G, Zulli C, et al. Pycnogenol® supplementation improves cognitive function, attention and mental performance in students. Panmin-erva Med 53: 75–82, 2011

18. Belcaro G, Luzzi R, Dugall M, Ippolito E. Pycnogenol® improves cognitive function, attention, mental performance and specific professional skills in healthy professionals age 35–55. Minerva Med, 2014, submitted

19. Vinciguerra G, Belcaro G, Bonanni E, et al. Evaluation of the effects of supplementation with Pycnogenol® on fitness in normal subjects with the Army Physical Fitness Test and in performances of athletes in the 100-min-ute triathlon. J Sports Med Phys Fitness 53: 644–654, 2013

20.   Errichi S, Bottari A, Belcaro G, et al. Supplementation with Pycnogenol® improves signs and symptoms of menopausal transition. Panminerva Med 53: 65–70, 2011

21.   Yang H-M, Liao M-F, Zhu SY, et al. A randomized, double-blind, place-bo-controlled trial on the effect of Pycnogenol® on the climacteric syndrome in perimenopausal women. Acta Obstet Gynecol Scand 86: 978–985, 2007

22.   Belcaro G, Hu S, Cesarone MR, et al.A controlled study shows daily intake of 50mg of French Pine Bark Extract (Pycnogenol®) lowers plasma reactive oxygen metabolites in healthy smokers.Minerva Med 104: 439–446, 2013

Chapter Four | Healthy Bones and Joints

Perhaps the most interesting thing that we have found about Pycnogenol® over the decades is that we learn so much from those who are already taking it as a dietary supplement. Many people take Pycnogenol® for one specific health benefit and then unexpectedly find that another health condition also improves. We constantly receive such reports and are asked for explanations.

These "leads" have kept Professor Rohdewald busy verifying that it is indeed Pycnogenol® that's responsible for these unexpected health benefits. In addition to his basic research program designed to elucidate the biochemical parameters and mechanisms of the previously known health benefits, each new lead offers promising health benefits that have their own urgency and importance. The focus of the research led by Professor Rohdewald is to elucidate the nutritional pathways and mechanisms of the various nutrients in Pycnogenol® and determine how they help keep body chemistry within normal ranges. Results count most, but scientists want to know more. How does Pycnogenol® stimulate (up-regulate) genes to produce adequate amounts of the needed compounds and how does Pycnogenol® keep them within the desired ranges by neutralizing free radicals and down-regulating inflammation?

Bone and joint health is one example. Clinical trials prove that Pycnogenol® helps improve bone and joint health, but it has been only fairly recently that we have understood why.

Wear and Tear Results in Joint Pain

Our joints are subject to wear and tear and with increasing age the lining of joints, the cushioning cartilage, gradually degenerates. When cartilage has reached significant abrasion articular tissue will be affected and tissue trauma initiates a local inflammation. The consequence is a reduced flexibility of joints and predominantly pain. Inflammatory cells accelerate degeneration of joints by secreting reactive oxygen species ("oxidative burst"), pro-inflammatory cytokines and degenerative enzymes. This process is paralleled by increasing pain, which, left untreated, may reach excruciating levels.

Persons of any age can experience pain in their joints, but it especially affects athletes and senior citizens. It hurts the majority of people in the age over 65. Their joints become out of order. Mostly the hips or the knees, but also fingers or the spine are affected. Usually it is because there has been more wear and tear than the body can handle. This unhealthy condition can lead to osteoarthritis (OA). It's not just the pain; the mobility in general gets affected with joint stiffness and a grating sensation during movement. It affects interactions with family and friends and the whole community. Some people would rather stay home than take the burden of moving about.

Joint pain typically is a gradually developing problem most of us develop as we grow older. An occasionally painful joint will be taken care of with a pain pill and after a while, people will find themselves slowly but surely consuming an ever-increasing amount of them. Unfortunately, too many pain pills over a long time can cause undesired effects.

Cartilage

The joints between bones are cushioned by flexible connective tissue called cartilage. Healthy cartilage contains ample collagen protein and water, making it not as hard and rigid as bone, but stiffer and less flexible than muscle. Pycnogenol® helps nourish cartilage by its stimulation of collagen and hyaluronic acid production. (Ref. 1)

It is the aging of the cartilage that creates the problems. The older the cartilage, the lower the water content, the smaller the cartilage and the lower its elasticity. Finally, the cartilage does not function anymore as a "lubricant" or a "cushion" stopping the direct contact of the bones in the joint. Then, one bone tip rotates on the other bone tip, bone tissue gets damaged and the inflammation process is set into action. The NF-κB switch that we discussed in the previous chapter is at the core of deciding "go" or "no go" in determining whether or not immune cells shall be deployed to take care of suffering tissue, such as in the case of harmed cartilage.

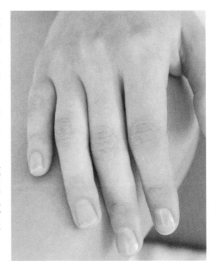

Human clinical trials show that with Pycnogenol® inflammatory cells release fewer enzymes that are harmful to the cartilage and also fewer COX enzymes responsible for the pain. Thus, cartilage experiences much less "friendly fire" from immune cells and has the opportunity to recover and rebuild. This process, of course, takes time and pain-lowering and improved flexibility with Pycnogenol® requires a little patience.

Pycnogenol® Is Different Than Prescribed Joint Health Medication

The more severe the joint pain gets, the more people depend on powerful painkillers. The most potent medications were the COX-2 specific inhibitors, which, unfortunately, cannot be taken continuously because of serious cardiovascular side effect risks. Pycnogenol® is a nutritional supplement. Pycnogenol® is not meant to replace a painkiller and it doesn't work in the way as they do. With the painkillers, you may expect significant pain alleviation within 30–60 minutes. But beyond the pain reduction, the analgesic medication does nothing to improve your joints. On the contrary, the disrupted pain sensation may cause people to further over-use the joints and this leads to an ever-increasing reliance on pain medication.

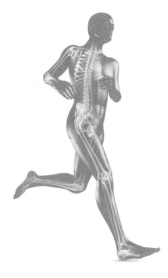

There are often undesired side effects from long-term medication with acetaminophen (e.g., decreased levels of the master antioxidant glutathione) and aspirin (e.g., internal bleeding). Being able to wean off medication and take a healthy nutrient instead has double benefit. Because of its low rate of side effects and its manifold anti-inflammatory properties, Pycnogenol® is an excellent adjunct for reducing joint pain.

Pycnogenol® may take weeks to produce noticeable pain relief, however, here the pain relief results from a better joint cartilage condition. Pycnogenol® aims at the inflammatory processes prevailing in the joint and calms down the immune cells, stopping them from taking the cartilage under "friendly fire." This is the prerequisite for having the cartilage recover and this process takes time. The three clinical trials done with Pycnogenol® so far all show that patients gradually require fewer painkillers to live with their arthritic joints. The most recent study actually showed that patients then also experienced much less stomach problems, a result of years of chronic pain killer usage.

Iranian Study

In a clinical trial conducted in Iran, 37 patients with osteoarthritis received either placebo (inert pill) or 150 mg Pycnogenol® in a double-blind study. (Ref. 2) A double-blind study is one in which neither the researchers nor the subjects know which pill is the placebo pill or the active pill. This reduces unintentional bias. In the study, the patients reported in monthly intervals their use of "pain killers" and their symptoms. The patients in the placebo group needed the pain killers on more days than before, increasing the dose during the treatment period, whereas the Pycnogenol® group could miss the analgesic treatment progressively more and more. Despite the higher use of anti-inflammatory drugs, the placebo group reported no significant improvement of symptoms. On the contrary, the participants of

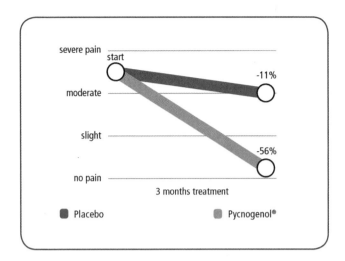

Figure 4.1. Results from the Italian clinical trial with Pycnogenol® for 156 osteoarthritis patients.

the Pycnogenol® group reported a continuous decline of symptoms, ending on the third month with 43% less pain, 35% less stiffness and 52% improvement of physical function of the joints. If one takes into account the low intake of pain killers, a very promising result of this first trial.

Slovakian Study

In the next study, 100 osteoarthritis patients were included in a double-blind, placebo-controlled investigation, performed in a Department of Orthopedics in Slovakia. (Ref. 3) Results were in complete agreement with the Iran study: The patients required less additional medication in the Pycnogenol® group, symptoms of osteoarthritis improved from month to month relative to start and placebo. Two weeks after the end of the study there was no significant relapse of pain and symptoms in the Pycnogenol® group, pointing to a persisting effect of the anti-inflammatory activity of 150 mg Pycnogenol®.

Italian Study

The third study involved 156 patients and was conducted in Italy. This was a double-blind, placebo-controlled study, over three months. (Ref. 4)

Table 4.1. Change of osteoarthritis scores (WOMAC)

Symptoms and functions	Pycnogenol® group		Placebo group	
	Inclusion	3 months	Inclusion	3 months
Pain Walking	3.3	2.1	3.0	3.0
Pain Stair Climbing	3.2	1.2	3.3	3.1
Pain Weight bearing	4.2	2.2	4.1	3.6
Morning stiffness	3.4	2.1	3.3	3.6
Stiffness during the day	3.2	1.1	3.4	3.1
Descending stairs	3.3	1.4	3.1	2.5
Ascending stairs	3.2	0.5	3.3	2.6
Rising from sitting	3.1	0.8	3.0	3.1
Getting in or out of car	4.0	1.2	3.8	3.0
Putting on socks	3.6	2.1	3.4	3.2
Rising from bed	3.5	2.2	3.2	3.0
Getting on or off toilet	3.8	2.1	3.5	3.0
Heavy house duties	3.8	2.0	3.2	2.8
Light domestic duties	3.6	1.1	3.5	3.0
Leisure activities	3.4	2.2	3.4	3.2
Community events	3.3	2.0	2.8	3.0
Church attendance	3.4	1.1	3.2	3.1
Leisure activities with friends	3.2	1.0	3.0	2.6
Leisure activities with others	3.2	1.1	2.9	3.1
Anxiety	3.4	0.3	3.2	2.5
Frustration	3.3	1.3	2.9	3.0
Depression	2.2	1.0	3.0	2.1
Insomnia	2.5	0.4	2.1	2.1
Boredom	3.4	0.7	2.8	2.2
Well-being	3.6	1.1	3.5	3.0

In this study, special emphasis was given to the relevance of Pycnogenol® to daily life. Of course, for the patients, the most needed benefit is pain relief. With 100 mg Pycnogenol® daily, pain decreased by 52% and concomitant medication was lowered by 58%. Under the placebo, pain decreased by just 11%, while concomitant medication increased by 1%. The stiffness score was halved with Pycnogenol® and the joint physical function improved by more than 50%. In the control, joint stiffness remained unchanged and only a marginal improvement of physical function was seen.

These great improvements were reflected in daily life. The patients were asked to rate the development of their social functions and emotions in parallel to their symptoms. The disease, due to the limited mobility, restricts many aspects of daily life.

Because of the striking amelioration of symptoms, the Pycnogenol® patients were less stiff, more mobile. So they could now participate more in community events, could go to church, and could visit friends. This resulted mentally in less frustration, less depression. The global score for limited social activities and emotional impact dropped significantly from 31.4 to 11.5. This sophisticated approach of this study in examining the improvement in real life – and not just the markers of inflammation – gives patients very worthwhile information.

Walking greater distances, less edema

Another test was an exercise having the objective of measuring the progress in mobility. With 100 mg Pycnogenol®, patients could extend their walking distance nearly threefold.

Table 4.2. Exercise test before and after 3 months

	Pycnogenol®	Placebo
Inclusion	68 m	65 m
3 months	198 m	88 m

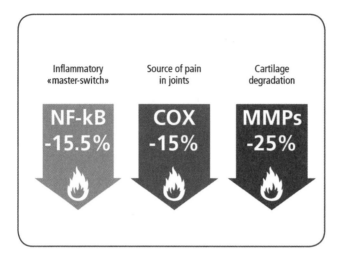

Figure 4.2. Pycnogenol® inhibits inflammation mediators.

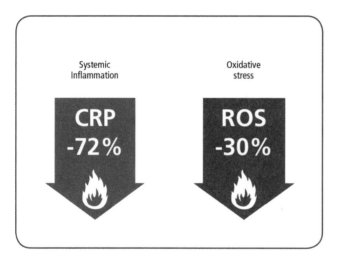

Figure 4.3. Pycnogenol® lowers the inflammatory marker CRP in osteoarthritis patients.

Immobility hinders the circulation in the lower limbs. More than 70% of the patients showed visible foot edema before the start of the study. At the end of treatment, edema (swelling due to the accumulation of fluid) decreased in 79% of patients in the Pycnogenol® group, but only in 1% in the placebo group. As a more comprehensive measurement, hospital admissions dropped down by 60% in the Pycnogenol® group, versus 4% in the control group.

The reason for all these positive results is based primarily on the manifold anti-inflammatory effects of Pycnogenol®. Patients with osteoarthritis produce a lot of reactive oxygen species (ROS) and inflammatory marker CRP, which are both significantly reduced under Pycnogenol®.

Also important is the inhibition of the protein-destroying enzymes (MMP) by 25%, the inhibition of the pain-producing prostaglandins by 15% and the inhibition of the master switch of inflammation, NF κB by 15.5%, which we discussed in Chapter Three.

In addition, rebuilding of the cartilage by the increased synthesis of the "lubricant" hyaluronic acid and of collagen with Pycnogenol® (Ref. 1), should help the patients to recover.

Less pain killers, less problems with the stomach

Unintended effects were reported in diaries of the patients. Gastrointestinal troubles decreased by 63% in the Pycnogenol® group, but only by 3% in the placebo group. (Ref. 4) It is reasonable to assume that the decrease in intake of non-steroidal analgesics decreased in turn the gastric problems in the Pycnogenol® group.

From these studies, one can conclude that Pycnogenol® helps maintain joint health and allows a sparing in the use of pain killers, thereby reduces their side effects. Pycnogenol® enhances the quality of life of osteoarthritic patients by relief of symptoms, leading to better mobility.

Table 4.3.

Overview of three clinical trials demonstrating efficacy of Pycnogenol® for arthritis

| Study | Number of patients | Arthritis symptom relief after 3 months Pycnogenol® relative to baseline (* after 2 months) | | |
		Pain	Joint stiffness	Physical function
Farid et al.	37	- 43 %	- 35 %	+ 52 %
Cisar et al.	100	- 40 %	- 40 %*	+ 22 %*
Belcaro et al.	156	- 55 %	- 53 %	+ 56 %

These studies demonstrate that Pycnogenol® promotes joint mobility and flexibility and naturally relieves the aching. They also show that after supplementing with Pycnogenol®, joints were more flexible and less medication was required. In addition Pycnogenol® has been found to lower inflammatory markers of joint soreness.

These bone and joint health benefits are brought about mainly by Pycnogenol®'s anti-inflammatory and antioxidant actions, as well as its effect on bone and cartilage collagen. Now, let's look at the role of Pycnogenol® in the Protection Against Metabolic Syndrome.

Chapter References

1.   Marini A, Grether-Beck S, Jaenicke T, et al. Pycnogenol® Effects on Skin Elasticity and Hydration Coincide with Increased Gene Expressions of Collagen Type I and Hyaluronic Acid Synthase in Women. Skin Pharmacol Physiol 25:86–92, 2012

2.   Farid R, Mirfeizi Z, Mirheidari M, et al. Pycnogenol® supplementation reduces pain and stiffness and improves physical function in adults with knee osteoarthritis. Nutrition Res 27: 692–697, 2007

3.   Cisár P, Jány R, Waczulíková, et al. Effect of Pine Bark Extract (Pycnogenol®) on Symptoms of Knee Osteoarthritis. Phytother Res 22: 1087–1092, 2008. Belcaro G, Cesarone MR, Errichi S, et al.

4.   Treatment of Osteoarthritis with Pycnogenol®. The SVOS (San Valentino Osteoarthrosis Study). Evaluation of Signs, Symptoms, Physical Performance and Vascular Aspects. Phytother Res 22: 518–523, 2008

Chapter Five | Protection Against Metabolic Syndrome

The bad news is that modern diets and lifestyles have created the two closely related pandemics of "Metabolic Syndrome" and diabetes. The good news is that Pycnogenol® reduces the risk of both by helping to maintain glucose (blood sugar) and insulin levels in the normal range. Metabolic Syndrome is often called Prediabetes. Similar biochemical pathways apply to both. We will discuss Metabolic Syndrome in this chapter and diabetes in the following chapter.

Although Pycnogenol® is not a pharmaceutical that is intended to treat diabetes, it is an important nutritional adjunct that many physicians recommend to help diabetics in many ways. As examples, studies show that Pycnogenol® can:

- improve all five traits of the Metabolic Syndrome,
- help maintain blood sugar (glucose) levels in the normal range,
- slow digestion of refined carbohydrates and sugars,
- reduce blood triglyceride levels,
- improve blood HDL (the so-called "good" cholesterol) levels,
- reduce the dangerous free radicals caused by diabetes,
- protect the eyes against diabetic cataract,
- protect the eyes against diabetic retinopathy,
- protect the nerves against diabetic neuropathy,
- protect the kidneys against diabetic nephropathy,
- help keep blood pressure within the normal range, and,
- help heal diabetic ulcers.

Metabolic Syndrome

Metabolic Syndrome was originally called "Syndrome X" and it is occasionally referred to as "Prediabetes Syndrome." Metabolic Syndrome is defined as a cluster of two or more of the following traits: abdominal obesity (paunch or beer belly), high blood pressure, low HDL (good choles-terol), high triglycerides, and insulin resistance. The more of these traits a person has, the greater the risk of cardiovascular disease and diabe-tes. Insulin resistance – where the body's cells have become inefficient in using insulin – is the cornerstone of Metabolic Syndrome and it leads to all of these other problems. Please refer to Box 5.1 for more details.

Box 5.1 Metabolic Syndrome Definition.

The Metabolic Syndrome is defined by the International Diabe-tes Federation by the following criteria:

- Waist circumference: greater than 94 cm for European men and greater than 80 cm for European women.

- Elevated triglycerides: greater than 150 mg/dL.

- Reduced "good cholesterol:" HDL less than 40mg/dL for males and less than 50mg/dL for females.

- High blood pressure: Systolic (upper number) greater than 130 or diastolic (bottom number) greater than 85 mm Hg.

- Fasting glucose: greater than 100mg/dL.

Dietary sugars and other refined carbohydrates trigger a rapid increase in blood glucose levels. In response, the pancreas quickly releases large amounts of insulin. High blood insulin levels eventually overwhelm the insulin receptors on cells, and they become insulin resistant. In other words, cells cannot properly use insulin to "burn" sugar for energy. As a result, more and more of the glucose is converted to fat, and some glucose remains in the blood, leading eventually to type 2 diabetes.

Obesity contributes to the development of type 2 diabetes and its pre-cursor insulin resistance. Obesity is associated with low-grade chronic inflammation characterized by inflamed adipose tissue with increased macrophage (white blood cells) infiltration. This inflammation is now widely believed to be the key link between obesity and development of insulin resistance. (Ref. 1)

Insulin resistance can develop when macrophages, clustered inside fat tissue, secrete inflammatory mediators damaging the surrounding cells. This renders them unable to control blood glucose levels in response to the insulin signal. A large mass of fatty tissue produces more inflammation so more cells become resistant to insulin.

Strategies against Metabolic Syndrome include dietary and lifestyle interventions as well as medications. Most medications are intended to reduce the risk of heart disease and diabetes.

Frequency of the Metabolic Syndrome

The Metabolic Syndrome is a world-wide pandemic. The National Health Statistic reported 2009 that a little more than one-third of the adults in the USA fulfill the criteria for having a Metabolic Syndrome: 35.1% of males and 32.6% of females.

From 1988–1994, the total number of persons in the USA with Metabolic Syndrome was 29% of the adult population, increasing to 34% between 1999–2006.

The statistics for Canada and Europe approximate 25% for large cities, in Latin America 14–27%, Asia between 10 and 20%, Japan just 5–8%. Despite ethnic differences, the trend towards higher prevalence of the Metabolic Syndrome is global.

The Metabolic Syndrome represents an enormous burden for the health care systems worldwide. The burden increases every year in every developed country. It's a human trait. People like to sit or "hang around" instead of walking, they like to eat fatty and sweet stuff, some smoke, and some enjoy alcoholic drinks. It is the sedentary, fast food lifestyle causing the Metabolic Syndrome that increases the risk of stroke, heart attack and diabetic complications.

It seems to be ridiculously easy to avoid the Metabolic Syndrome: eat less food, eat healthier food, and exercise moderately. That's it. But real world experience tells us that it is extremely difficult to change a person's lifestyle. It is nearly utopic to change the lifestyle of the modern urban populations. Even in a population adhering to the healthier Mediterranean diet one can identify citizens having the Metabolic Syndrome, for example in the village of San Valentino in Central Italy.

Table 5.1 Decrease of symptoms of Metabolic Syndrome.

		Inclusion	3 Months		6 Months	
		Pycnogenol® N=76	Control N=78	Pycnogenol® N=64	Control N=66	Pycnogenol® N=64	Control N=66
Waist circumference	Male	106.4	105.8	98.8	101.3	98.3	100.2
	Female	90.9	91.2	84.6	89.2	83.6	87.3
Triglycerides		189.3	194.3	143.23	179.4	143.2	182.3
HDL cholesterol	Male	35.4	35.9	41.2	37.4	44.5	39.3
	Female	44.5	45.5	53.3	45.9	55.4	46.1
Blood pressure	Systolic	144	143.2	138.2	141.3	137.4	142.3
	Diastolic	87.6	87.2	82.4	83.4	82.3	85.2
Fasting glucose		123.2	124.3	106.4	118.3	105.3	119.3

San Valentino, Italy, Six-Month Study

Residents of San Valentino, in Central Italy, presenting the Metabolic Syndrome, were recruited for a study to see how the nutritional benefits of Pycnogenol® would help them. (Ref. 2). A group of 76 volunteers who had all five symptoms of the Metabolic Syndrome was given 150 mg of Pycnogenol® daily and studied for six months. They were compared with a control group of 78 patients. They received the same instructions as the Pycnogenol® group. As examples, all participants were moderately active, used stairs instead of elevators, walked to shop rather than using a car when possible, ate whole grain pasta and consumed no ready-made meals.

The participants were checked at the start of the study and then again after three months and finally after six months for all five symptoms of the Metabolic Syndrome. Members of the control group noted a slight improvement in the majority of symptoms, most probably as a result of a moderate change in lifestyle. However, the changes failed to reach significance levels after 3 months.

After six months, all five symptoms were still present in 35 persons in the control group, but in only one person in the Pycnogenol® group. Seven volunteers ended free of each of the five symptoms in the Pycnogenol® group, but none in the control group did so.

Statistical comparison of the changes in the Pycnogenol® group revealed a significant amelioration of every single symptom in just three months. The values after six months were almost the same or slightly better. Blood glucose dropped by 14.4%, waist circumference dropped 3.11 inches (7.4%) in men and 2.9 inches (8%) in women. Also, plasma free radicals dropped by 24.6% in the Pycnogenol® group.

Just the Nutritional Supplement Pycnogenol®

These results demonstrate that Pycnogenol® was able to reduce waist circumference, to lower triglycerides, to increase the "good" cholesterol (HDL), to lower blood pressure and to lower blood glucose. The mean values of the Pycnogenol® group reached the normal values, with only the systolic (upper number) blood pressure slightly over the normal level.

The participants had not taken an antihypertensive drug plus a cholesterol-lowering statin plus a triglyceride-lowering fibrate plus an anti-diabetic drug. Just one type of pill was needed – just the nutritional supplement Pycnogenol®.

This investigation verified that Pycnogenol® was able to exert its manyfold health benefits in those who really need this simultaneous anti-diabetic and antihypertensive action, together with an improvement of blood lipids. The significant reduction of waist circumference may depend on the combined effects of a normalized metabolism plus a moderate change in lifestyle according to the study protocol.

It should be emphasized that although it was not a part of this study, Pycnogenol® also reduced the risk of the volunteers against deadly blood clots (thrombosis), a frequent risk for diabetic and hypertensive persons. As we discussed in Chapter Two on Heart Health, this nutritional action of Pycnogenol® has been well-documented in other studies. We also discussed that while Pycnogenol® has the same anti-clotting benefit as aspirin, Pycnogenol® does so without increasing the risk of intestinal bleeding.

All the Health Benefits Corroborated in One Study

This Italian study showed that Pycnogenol® did in reality what could be expected theoretically. Its actions against diabetes, hypertension, and high blood cholesterol were already known. The finding of a normalization of high triglycerides by Pycnogenol® was new, however. The observed reduction of waist circumference is undoubtedly highly desirable, however, the contribution of moderate lifestyle changes shouldn't be neglected.

As we mentioned in earlier chapters, sometimes we learn of additional health benefits of Pycnogenol® from those taking the nutritional supplement for one health benefit or another and then finding that their health also improved in additional and unexpected ways. Such anecdotal reports must be followed with formal studies to determine if the observed health benefits are real. If they are scientifically verified, then many other people can also benefit from this information. As you'll read in Chapter Six, such is the experience with Diabetes.

Chapter References

1.   Oliver E, McGillicuddy F, Phillips C, et al. The role of inflammation and maceophage accumulation in the development of obesity-induced type 2 diabetes mellitus and the possible therapeutic effects of long-chain n-3 PUFA. Proceedings of the Nutrition Society, 69:232–243, 2010

2.   Belcaro G, Cornelli U, Luzzi R, et al. Pycnogeno® supplementation improves health risk factors in subjects with metabolic syndrome. Phytother Res 27:1572–1578, 2013

Chapter Six | Pycnogenol®
Against Diabetes

The U. S. Centers for Disease Control and Prevention (CDC) describes diabetes mellitus as a disease in which glucose (blood sugar) levels are above normal. Most of the food we eat is turned into glucose, a simple sugar, for our bodies to be used for energy production. The hormone insulin helps glucose to get into the cells. When someone has diabetes, his or her body either doesn't make enough insulin (type 1 diabetes) or can't use its own insulin as well as it should (type 2 diabetes). This causes sugar to build up in the bloodstream and has other negative effects such as elevating blood lipids and damaging blood vessels—all of which can cause a number of serious health problems.

The damage to the cells that leads to either type 1 (juvenile) diabetes or type 2 (adult onset) diabetes usually involves free radical reactions, but once the islets of Langerhans cells of the pancreas (type 1) or cellular mechanisms for utilizing insulin (type 2) have been damaged, antioxidants cannot reverse this. However, diabetes itself increases the production of free radicals, which further damage the body thus increasing the risk of heart attack, nerve damage (diabetic neuropathy), cataract (diabetic cataract), blindness (diabetic retinopathy), kidney damage (diabetic nephrology) and more complications. Here's where powerful antioxidant protection of Pycnogenol® is especially important – diabetics need more antioxidant protection than healthy persons.

Types of Diabetes

The three main types of diabetes are type 1 diabetes, formerly called insulin-dependent diabetes mellitus or juvenile-onset diabetes; type 2 diabetes, formerly called non-insulin-dependent diabetes mellitus or adult-onset diabetes; and gestational diabetes that only pregnant women get.

Type 1 diabetes is an autoimmune disease in which the immune system attacks the insulin-producing beta cells in the pancreas and destroys them. The pancreas then produces little or no insulin, which requires the diabetic to take insulin daily. This disorder, which generally develops in children and young adults (although it can occur at any age), accounts for about 5 percent of all diagnosed cases of diabetes.

Type 2 diabetes results when the body gradually loses its ability to use and produce enough insulin. This disorder typically appears in adults but is increasingly being diagnosed in children and adolescents as maturity-onset diabetes of the young (MODY). It is usually part of a Metabolic Syndrome that includes excess weight/obesity, insulin resistance, high triglycerides/low high-density lipoprotein (HDL often called "good") cholesterol, and/or high blood pressure. It accounts for about 90 to 95 percent of all diagnosed cases of diabetes.

Gestational diabetes is a type of diabetes that only pregnant women get. If not treated, it can cause problems for mothers and babies. Gestational diabetes develops in two to ten percent of all pregnancies but usually disappears when a pregnancy is over.

Other specific types of diabetes resulting from specific genetic syndromes, surgery, drugs, malnutrition, infections, and other illnesses may account for one to five percent of all diagnosed cases of diabetes. In this chapter, we'll focus on type 1 and type 2 diabetes.

Help Prevent Inflammation Associated with Obesity

The increasing prevalence of overweight and obesity is a major contributing factor to the epidemic of type 2 diabetes. The World Health Organization (WHO) estimates that more than 1 million deaths annually in Europe can be attributed to diseases related to excess body weight; in the United States, this estimate is 300,000, according to a recent U.S. Surgeon General's report. And with the rising global obesity levels, these death rates are set to drastically increase. Roughly 80 percent of those with type 2 diabetes are overweight.

We Listen to Pycnogenol® Users

Anecdotal reports can be interesting "case studies of one," but they are not proof of anything. However, they often suggest subjects for further scientific study. Professor Rohdewald remembers the three anecdotal reports that prompted his formal research into the nutritional actions of Pycnogenol® and insulin biochemistry. The first hint came several years ago when a caller asked, "Where can I buy Pycnogenol®? It was so helpful against my diabetes. I got it from my daughter in the US." He explained to Professor Rohdewald, "Now, I no longer need insulin and I don't have to go to the hospital emergency room that often."

As Professor Rohdewald relates, "I was a bit shocked. I realized that this man is a patient with an uncontrolled diabetes, who needs definitely insulin. I had never heard before that Pycnogenol® was able to lower blood sugar in diabetics. Pycnogenol® is a nutritional supplement, not a pharmaceutical." He continues, "So I gave him the strict advice to stick to the orders of his physician and to take insulin at a more regular basis. And I explained him, to his disappointment, that there is nothing known about benefits of Pycnogenol® as an anti-diabetic."

We did, however, have results that Pycnogenol® had been helpful against diabetic retinopathy. This beneficial effect, however, was not specific for diabetes as Pycnogenol® also improved eye health in case of other retinopathies as well. The reason for Pycnogenol®'s effects on eye health, the sealing of leaky capillaries inside the eyes and improved microcirculation, gave no hint for an anti-diabetic effect.

"My father doesn't need insulin anymore," part 2

Just a few months later while Professor Rohdewald was lecturing in South Korea, he was approached after a lecture by a member of the audience mentioning that his father has benefited from Pycnogenol®. As Professor Rohdewald remembers, "remarkably, he explained to me that his father no longer needs insulin because he takes Pycnogenol® each day and his blood sugar is under control."

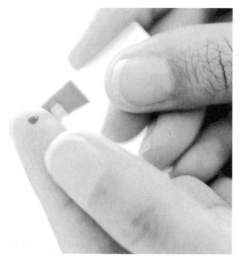

As Professor Rohdewald continues, "Having the phone call from the German diabetic patient in mind, I asked him in detail about the situation of his father. At the very end, I was convinced that the father had diabetes type 2 indeed and, according to the son's report – the father used Pycnogenol® as an anti-diabetic food supplement. Of course I tried to understand why, but I couldn't imagine a mechanism how the extract could bring the blood sugar down. I came to the conclusion that it was probably the effect of a better dietary control under the guidance of the son."

"I no longer need insulin," part 3

However, in the following week, it happened again in Japan. This time it was a woman who approached Professor Rohdewald after his lecture. With the aid of his translator, Professor Rohdewald learned that the woman claimed she was a diabetic who no longer needed insulin. Further interrogation by Professor Rohdewald determined that she had type 2 diabetes, was on insulin, didn't change her diet, then took Pycnogenol® as a nutritional supplement and after her blood glucose now routinely tested okay, she was able to gradually reduce her insulin and eventually do away with it.

Could three such anecdotal reports in such a short time be a coincidence? Or was there a real effect of the nutrients in Pycnogenol® in maintaining blood glucose with in normal limits? These anecdotal reports required a research decision.

Three Anecdotal Reports – A Reason for Proof

Professor Rohdewald just had to find out what, if anything, was going on with the nutrients in Pycnogenol® and blood glucose. "Normally, a scientist develops a hypothesis and then starts tests to find out whether the hypothesis is true or not," he reminds us. "But in this case, I had no idea why the Pycnogenol® would help keep blood glucose within normal limits. I had to test whether these three reports from three countries with the same tenor reflected actually a previously unknown effect of Pycnogenol® against diabetes. In addition, reports came on my desk from Australia that feeding diabetic rats with Pycnogenol® lowered their blood sugar. (Ref. 1) This was finally the last reason to conduct formal clinical studies."

First proof: Pycnogenol® lowers blood sugar dose-dependently

The first clinical studies of Pycnogenol® and blood glucose were made in Peoples Republic of China. To determine whether there is a true effect against diabetes one has to find out whether blood glucose in diabetic patients is lowered following intake of Pycnogenol®. Furthermore, a higher dose of Pycnogenol® should affect blood glucose more than a lower dose. If the effect is dependent on the dose, there is a good reason to assume that there is a true effect.

The tests were successful: In 30 volunteers with type 2 diabetes responded better with higher intakes of Pycnogenol® – 200, 300 mg – better than to 50 and 100 mg. The plain message was: The higher the dose, the lower the blood sugar. (Ref. 2)

According to a clinical study conducted under the leadership of Dr. Lau and published in *Diabetes Care*, scientists discovered that type II diabetes patients had lower blood sugar and healthier blood vessels after supplementing with Pycnogenol®. (Ref. 2)

The open, controlled, dose-finding study demonstrated that patients with mild type II diabetes, subscribing to a regular diet and exercise program, were able to significantly lower their blood sugar levels when they supplemented with Pycnogenol®. A dosage as low as 50 mg significantly lowered blood sugar. 100 mg dose lowered blood sugar levels even more, whereas higher dosages only marginally further increased the effect.

But the question was now:
Why was the blood sugar lowered by Pycnogenol®?

One possible mechanism of action could be excluded: Pycnogenol® did not lower blood sugar because of a stimulation of insulin secretion. Insulin levels had been tested in this study (Ref. 2) and remained unaffected by Pycnogenol®. So we had still to look for another explanation.

Knowing from this dose-finding study, that the anti-diabetic effect was science, not fiction, we started a new investigation in China.

Second proof: the first double-blind study

This time, we used the accepted "gold standard" of clinical sciences, a randomized double-blind study, placebo-controlled. Randomized means that the volunteers are assigned to the various groups at random. Double-blind means that neither the volunteer nor the researcher knows who is getting what until after the results are tabulated. A placebo is merely an inert pill that looks like the test pill. This study, published in the October 2004 issue of *Life Sciences* (Ref. 3), found that 77 type 2 diabetes patients who supplemented with 100 mg of Pycnogenol® for 12 weeks, during which a standard anti-diabetic treatment was continued, significantly lowered blood glucose levels as compared to a placebo.

Simultaneously, other biomarkers indicated in this study a protection against thrombosis and improved cardiovascular function in the Pycnogenol® group. Nitric oxide and prostacyclin, vasodilating substances were

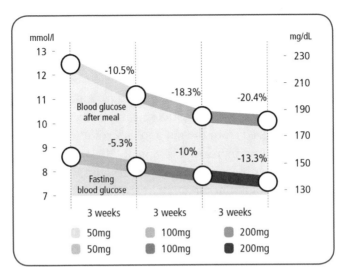

Figure 6.1. Pycnogenol® lowers blood glucose (blood sugar).

found in larger quantities in plasma, whereas the potent vasoconstrictor endothelin 1 diminished.

Third proof: the second double-blind study

A similar investigation in the U.S. with 48 diabetic patients, suffering in addition from elevated blood pressure, confirmed these findings. (Ref. 4) Patients taking 125 mg/day Pycnogenol® lowered their blood sugar and could control their high blood pressure with less anti-hypertensive drugs. In addition, the "bad" cholesterol (LDL) diminished during treatment with Pycnogenol®.

While the results of the clinical investigations with 155 patients confirmed the initial anecdotal reports, Professor Rohdewald still could not answer the question of how Pycnogenol® works to normalize blood glucose levels. One could speculate that the anti-oxidative properties of the pine bark could be the reason, but this was not really convincing.

Answer to the Question: How?

Experiments made at a university in Germany offered finally an explanation. Pycnogenol® blocks an enzyme (glucosidase) that is needed for
splitting (digesting) of carbohydrates to produce glucose. (Ref. 5) If carbohydrates are digested less, less glucose is formed. Then less glucose is
available to be absorbed from the gut and as a result the glucose concentrations in blood cannot increase. Pycnogenol® delays the uptake of sugar
from a meal 190 times more potently than prescription medications, preventing the typical high glucose peak in the blood stream after a meal.
(Ref. 5)

Another finding, in cell culture, showed that cells could absorb glucose
better from blood under the influence of Pycnogenol®. (Ref. 6) Without
insulin, these cells are not able to extract glucose from blood. Pycnoge-
nol® stimulates cells for uptake of glucose even in absence of insulin, so
that glucose travels from blood into the tissue and blood sugar goes down.

At the end of the day, it was true that Pycnogenol® acts against high blood
sugar and there was some evidence why.

Consequences of Diabetes

However, the abnormal glucose level in blood is only a symptom of diabetes. The disease has widespread, serious consequences on health. The
lack of insulin disturbs every pathway of our metabolism, which in turn
harms eyes, blood vessels, kidneys. Such damage may lead ultimately
to blindness, hypertension, thrombosis, stroke, ulcers of the lower legs,
maybe amputation. Therefore, patients with diabetes are at high risk for
several serious damages of their organs.

Hence the question arose whether Pycnogenol® could help to prevent
these late consequences of diabetes type 2, in addition to lowering blood
glucose.

Hypertension and Thrombosis in Diabetes Patients

We were already happy to see in our investigations with diabetic patients that Pycnogenol® lowered blood pressure. And it changed some bio-chemical indicators into the direction of thrombosis prevention. These findings have been already discussed in detail in Chapter Two on Heart Health and suggest that Pycnogenol® provides a protection for the vascular system of diabetic patients. This is a very important point, as patients with diabetes die more often from thrombosis or stroke than the normal population.

Diabetic Retinopathy

Another serious consequence of diabetes is retinopathy, the leading cause of blindness in the USA. More than half the people suffering from diabetes may eventually develop retinopathy, an eye disease caused by capillaries leaking blood into the retina that destroys the light-sensing cells and leads to gradual vision loss.

Fundamentally, it is the permanently high glucose concentration in blood that causes brittle and leaky small blood vessels in the eyes. In the first stage, blood flows out from the leaky blood vessels and fluid, lipids and blood disturb eye sight. In a more advanced stage of the disease, blood capillaries are blocked, so that the surrounding tissue deceases.

Consequently, a tight control of blood glucose is needed to keep the glucose level as normal as possible. In addition to all efforts to keep the glucose level under control, the strengthening of the capillaries is the next target in prevention.

The natural polyphenol nutrients in Pycnogenol® help maintain the proper permeability of the walls of blood capillaries. When walls of blood capillaries become too permeable, as in case of scurvy or diabetic retinopathy, blood may pass the capillary wall and bleeding occurs, so that the escaped blood cells and lipids from plasma disturb eyesight. Pycnogenol® tightens blood vessel walls to the proper permeability, as has been shown

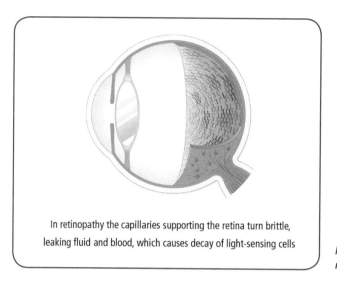

In retinopathy the capillaries supporting the retina turn brittle, leaking fluid and blood, which causes decay of light-sensing cells

Figure 6.2. Diabetic retinopathy.

in many investigations. Therefore, the leakage of retinal capillaries is significantly reduced following intake of Pycnogenol®.

Six clinical studies with more than 1,200 diabetics have shown that Pycnogenol® is effective in strengthening retinal capillaries, stopping further progression of retinopathy and saving the eyesight of diabetics.

Better Visual Acuity with Pycnogenol®

As a consequence of restoring the proper permeability to the capillaries in the eye, Pycnogenol® improved visual acuity in patients with diabetic retinopathy in several investigations in France and Germany.

Diabetic Retinopathy is considered a "stealth disease" as it progresses unnoticed and symptom-free while it leads to gradual, largely irreversible loss of vision. Left untreated retinopathy may progress to the proliferative stage, characterized by growth of new capillaries to compensate for the lack of oxygen in the retina. These vessels grow uncontrolled and interfere with normal eye vision and, furthermore, tend to cause severe microbleedings. The proliferative stage of retinopathy may lead to complete blindness.

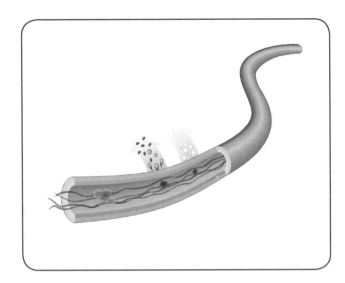

Figure 6.3. Pycnogenol® lends strength to retinal capillaries to decrease fluid leakage and control retinal microbleedings.

Pycnogenol® not only strengthens the retinal capillaries to help control leakage of fluids and blood into the retina, it also improves the endothelial function of retinal capillaries so that better blood microcirculation in the retina is obtained.

Two studies were carried out in France to explore the possibility of using Pycnogenol® as a nutritional adjunct for treatment of eye diseases resulting from capillary bleeding, mostly diabetic retinopathy. These studies were reported and published in French, the results have recently been reviewed in English. (Ref. 7)

In a double-blind comparative study, the efficacy of Pycnogenol® was compared with another compound commonly utilized for slowing the progression of diabetic retinopathy: calcium dobesilate (Dexium). In this German study, two groups with 16 diabetic retinopathy patients each were treated with either Pycnogenol® (120 mg/day for 6 days, then 80 mg/day) or Dexium (2–3 tablets equivalent to 1000–1500 mg calcium dobesilate per day) over a period of 6 months at the Ophthalmology Department of the University Clinic of Würzburg, Germany. (Ref. 7)

Particular emphasis was given in this study to obtain an objective judgment of the treatment efficacy. Panoramic photos of the entire retina were taken of all patients before and after treatment. In addition to studying

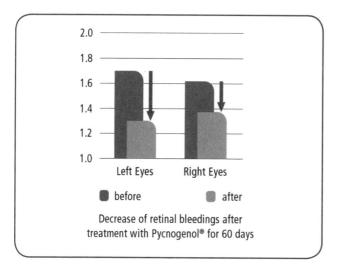

before   after

Decrease of retinal bleedings after
treatment with Pycnogenol® for 60 days

Figure 6.4.
Pycnogenol®
significantly lowers
retinal bleedings
after three months.

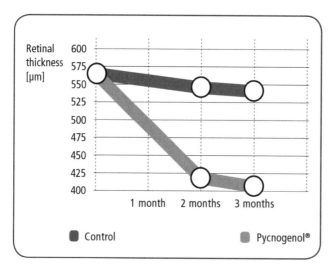

Figure 6.5.
Pycnogenol® lowers
retinal edema in
retinopathy. (Ref. 9)

the microbleedings, the study investigated exudates, which are the lipid depositions remaining in the retina from fluid leakages from capillaries. Seven ophthalmologists independently from one another judged the improvement of both bleedings and exudates, without knowing which medication the patient received. Both the retinal bleedings as well as the exudates improved in the majority of patients taking Pycnogenol®. The study outcome suggests a higher efficacy with Pycnogenol® as compared to Dexium.

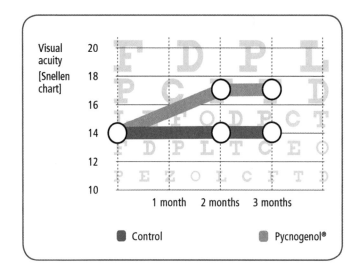

Figure 6.6. Pycnogenol® improves visual acuity in retinopathy. (Ref. 9)

Study Demonstrating Decreased Bleedings

In another clinical trial 40 retinopathy patients received intravenous injection of the dye fluorescein that aids in the identification and quantification of momentary retinal bleedings by measuring the fluorescence intensity of the dye. A rapid sequence of fluorangiographs allows for recording the retinal blood flow dynamics as well as the integrity of the blood-retina barrier. The microangiopathy was scored using a semi-quantitative 4-score scale ranging from healthy (=0) to severe bleedings (=3). After three months treatment with 150 mg/day Pycnogenol® retinal bleedings decreased significantly. (Ref. 8)

Pycnogenol® Improves Visual Acuity in Early Stage Retinopathy

A clinical study with 46 subjects suffering from early stage retinopathy characterized by mild to moderate retinal edema showed significantly improved visual acuity after three months treatment with 150 mg/day Pycnogenol®, whereas no effect was found in a control group. With Pycnogenol® the visual acuity improved on a Snellen chart from baseline 14/20 to 17/20. Moreover, this study demonstrated a significantly relieved retinal edema, a consequence of the increased capillary wall strength resulting from taking Pycnogenol® for three months. (Ref. 9)

The blood velocity was increased significantly by about 30% after taking Pycnogenol®, suggesting a better perfusion of retinal tissue, which is understood to be the reason for increased visual acuity. (Ref. 9)

Multi-Center Study with 1,169 Retinopathy Patients

The most impressive evidence for the efficacy of Pycnogenol® for saving the eyesight of retinopathy patients stems from a multi-center study in Germany. In this six-month study, 1,169 participants having either type 1 or type 2 diabetes, took Pycnogenol® in dosages ranging from 20 to 160 mg, depending on the severity of their retinal bleedings. The out-

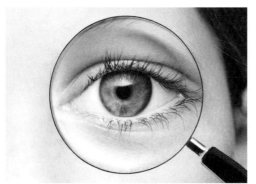

come of the study showed that after an average of six months, there was no further vision loss, demonstrating that Pycnogenol® effectively stops the progression of retinopathy. The study concluded, "Pycnogenol® has considerable therapeutic benefits for patients with diabetic retinopathy." (Ref. 7)

In summarizing the health benefits of Pycnogenol® in diabetic retinopathy, Pycnogenol® provides potent antioxidant protection against free radicals and oxidative-stress related degenerative processes in the eyes. The antioxidant properties of Pycnogenol® act in synergy with lipophilic antioxidants in the eye, such as with Lutein.

Pycnogenol®'s vascular benefits translate to considerable benefits for people suffering from retinopathy. This diabetic complication is improved with Pycnogenol® by nutritionally supporting impaired capillary integrity and function. The capillary filtration and microbleeding is significantly improved with Pycnogenol®. The better perfusion of retinal tissue owing

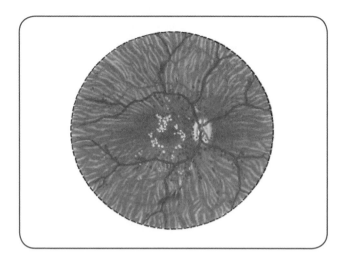

Fig 6.7 should appear at start of diabetic retinopathy

to improved endothelial function with Pycnogenol® helps restore some of the vision lost in retinopathy.

Kidney Damage

The deleterious influence of abnormally high, persisting glucose levels also damages the capillaries of the kidney. As with eyes, the kidney capillaries also become leaky. In the kidney, the result is that proteins escape from the capillaries from blood into urine. This loss of proteins with the urine progresses gradually over the duration of the diabetes. Again, this damage of the organ is the more severe the more the blood sugar is out of control. Under strict blood sugar control, the end of the damage of the kidney, kidney failure, may be avoided.

Better Renal Circulation, Less Loss of Proteins in Urine

As would be expected, intake of Pycnogenol® reduced the urinary excretion of proteins and enhanced kidney perfusion in a clinical investigation with 48 diabetic patients in the US (Ref. 4). Also in Italy, the disastrous loss of essential proteins could be reduced in 31 diabetic patients and also in 26 patients with high blood pressure. Measurements of renal circulation demonstrated a better perfusion of the kidneys (Ref. 10).

Table 6.2 Excreted albumin per day

	Control	Pycnogenol®
Inclusion	87 mg	91 mg
6 months	64 mg	39 mg
Decrease	26%	57%

Bad Circulation in the Lower Legs

Another serious complication of diabetes is the bad circulation inside the lower legs, leading to skin ulcers. The "open leg" may cause severe complications. We discuss this issue in Chapter 7 on Skin Health.

The Great Spectrum of Benefits for the Diabetic Patient

Taking all these results together one can conclude that the testimonials of the first consumers had a great impact on our research. Over a period of about 30 years our collected pieces of evidence suggest that Pycnogenol® is an excellent food supplement especially for patients with suspected or beginning, mild diabetes type 2.

Pycnogenol® is excellent not only because of its normalizing effect on blood sugar, but more because of its wide spread spectrum of preventive effects against the most serious complications of diabetes: hypertension, atherosclerosis, eye and kidney damage and ulcers of the lower legs.

Please keep in mind that diabetics are at increased risk for heart problems and that Pycnogenol® helps protect against heart disease as discussed earlier.

Chapter References

1.   Maritim A, Dene BA, Sander RA, et al. Effect of Pycnogenol® treatment on oxidative stress in streptozotocin-induced diabetic rats. J Biochem Mol Toxicol 17: 193–199, 2003

2.   Lau et al. French maritime pine bark extract Pycnogenol® dose-dependently lowers glucose in type II diabetic patients. Diabetes Care 27: 839, 2004

3.   Liu X, Wei J, Tan F, et al. Antidiabetic effect of Pycnogenol® French maritime pine bark extract in patients with diabetes type II. Life Sci, 75:2505–2513, 2004

4.   Zibadi S, Rohdewald P, Park D, et al. Reduction of cardiovascular risk factors in subjects with type 2 diabetes by Pycnogenol® supplementation. Nutr Res 28: 315–320, 2008

5.   Schäfer A, Högger P. Oligomeric procyanidins of French maritime pine bark extract (Pycnogenol®) effectively inhibit alpha-glucosidase. Diabetes Res Clin Pract 77: 41–46, 2007

6.   Lee HH, Kim K-J, Lee OH, et al. Effect of Pycnogenol® on glucose transport in mature 3T34-L1 adipocytes. Phytother Res 24: 1242–1249, 2010

7.   Schönlau F, Rohdewald P. Pycnogenol® for diabetic retinopathy: A review. Int Ophtalmol 24: 161–171, 2002

8.   Spadea L, Balestrazzi E Treatment of vascular retinopathy with Pycnogenol®. Phytother Res 15: 219–223, 2001

9.   Steigerwalt R, Belcaro G, Cesarone MR, et al. Pycnogenol® improves microcirculation, retinal edema, and visual acuity in early diabetic retinopathy. J Occul Pharmacol Ther 25: 537–540, 2009

10.   Stuard S, Belcaro G, Cesarone MR, et al. Kidney function in metabolic syndrome may be improved with Pycnogenol®. Panminerva Med 52: 27–32, 2010

Chapter Seven | More Than an Oral Cosmetic: Healthy and Beautiful Skin

Beautiful skin is not about vanity – it is important for overall health. Your skin is your first line of defense against the daily rigors of life. It's a vital organ that has several functions. Skin protects us from the environment, maintains hydration and body temperature, and is part of the immune system. Pycnogenol® has been called "the oral cosmetic" by many users because it gives skin a more vibrant glow. Pycnogenol® renews skin by rebuilding tissue, making skin more flexible and smoother resulting in healthier, younger-looking skin, with fewer fine lines and less discoloration. Supplementation with Pycnogenol® shows visible results, proven in clinical studies.

Pycnogenol® helps skin rebuild elasticity, which is essential for smooth and youthful looking skin. Pycnogenol® has a high and specific affinity for the main skin proteins, collagen and elastin. (Ref. 1) Like a "magic bullet," it finds its way to collagen and elastin, binds to them and protects them against free radicals and destructive enzymes. The most dangerous enzymes destroying the essential structure elements of the skin are the collagenases and elastases. Both enzymes are inhibited after intake of Pycnogenol®. (Ref. 1) Besides protecting collagen, a study has also shown that Pycnogenol® helps produce new fibers, which makes skin smoother and more elastic, with fewer wrinkles. (Ref. 2)

Pycnogenol® also stimulates the body's production of hyaluronic acid (Ref. 2), which is important for water retention, wound healing and filling in (volumizing) wrinkled skin to smooth it. This action counteracts the thinning of skin that develops with aging. Pycnogenol® helps the skin rebuild its thickness and elasticity. Skin fullness and elasticity are essential for skin smoothness.

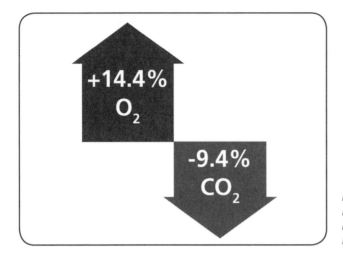

*Figure 7.1
Pycnogenol®
improves skin
respiration. (Ref. 3)*

Pycnogenol® Enhances Blood Micro-Circulation in the Skin

In addition to improving skin by its action with skin proteins, Pycnogenol® intake improves the microcirculation that hydrates, oxygenates and nourishes the skin. This improved microcirculation also clears toxins from the skin more quickly.

As we said, it's more than beauty, it's overall health. Let's look at the skin benefit of microcirculation improvement and how that not only produces healthier skin, but also can prevent ulcers in diabetics that otherwise cause amputations.

Pycnogenol® enhances generation of endothelial nitric oxide (NO) that is the key mediator facilitating arterial relaxation and consequently allows for optimal blood flow. Pycnogenol® was found to increase blood perfusion of the skin (Ref. 3), as could be shown by higher skin oxygen levels and, conversely, lower carbon dioxide concentration. This study demonstrated an improved healing of wounds (ulcers) in individuals with micro-circulatory disorders. An improved blood perfusion of the skin warrants optimal supply with all important nutrients as well as better hydration to support skin vitality. We'll discuss this more fully momentarily under the topic of "scars and ulcers."

Clinical Study of Skin Rejuvenation with Pycnogenol®

A study to provide some answers about Pycnogenol® and skin rejuvenation was performed at the University of Düsseldorf, Germany. (Ref. 2) In addition to determining Pycnogenol®'s effect on skin elasticity and smoothness, the study could gain information about how Pycnogenol® affected rejuvenation of skin at the molecular level. This study involved 20 volunteers in the "golden age" between 55 to 68 years of age.

To get the answers, the volunteers had biopsies taken from the part of the body needed for sitting. Skin in this area has been normally less damaged by the sun. Samples were taken at the start of the study and after 12 weeks with Pycnogenol®. The biopsies indicated significantly elevated levels of the messenger RNA (mRNA) that regulates (expresses) the enzyme hyaluronic acid synthase (HSA). This enzyme is needed for production of hyaluronic acid (HA). HA is an essential constituent of the dermal connective tissue, needed for elasticity and hydration of the skin. The substantial increase of 44% of HSA expression after supplementation with Pycnogenol® points to a beneficial effect on skin appearance. A higher expression (29–41%) of mRNA for collagen synthesis was also found in biopsies. This demonstrates a stimulation of collagen synthesis, the other important skin constituent.

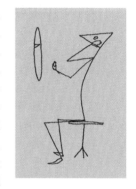

The next step in the study was to measure the cosmetic effects of Pycnogenol® using scientific methods. Skin hydration was measured by a corneometer, which registers the skin's electric properties. It was found that skin hydration was enhanced after 6 weeks; after 12 weeks hydration was lower, perhaps because of seasonal influence in summer. In a sub-group of 13 volunteers having dry skin the hydration was increased by 21%. (Ref. 2)

Skin elasticity was measured optically by a cutometer, which pulls on the skin by applying a vacuum and then releases the skin. The elasticity was significantly better after 6 and 12 weeks. Correspondingly, the fatigue of the skin was lower with Pycnogenol®.

Wrinkles were measured by using the Visioscan. A 3% reduction of wrinkles and an increase in skin smoothness was found.

Taking all data together, we have a mechanistic explanation for the improved skin elasticity and hydration because of enhanced synthesis of HA and collagen.

Additional Support

These results are in accordance with earlier findings that used a complex formulation that included Pycnogenol®. In a double-blind, placebo-controlled study with 62 women, skin elasticity was improved by 9% after the 6 weeks test period. (Ref. 4) Also in this investigation, a cutometer was used. Skin roughness, evaluated by three-dimensional microtopography imaging, was lowered by 6%.

To get a broader picture of the importance of Pycnogenol® to the skin, one has to add its manifold protective effects. Of utmost importance is that Pycnogenol® protects skin against UV radiation. Also important, just in context with elasticity, is the inhibition of the enzymes that destroy skin proteins – especially Pycnogenol®'s protection against elastase. This enzyme degrades the most important structural protein, elastin.

The synergism between protection of existing structures and creation of new structure elements favors Pycnogenol® for skin care.

Brown Spots, Hyper-pigmentation and Skin Whitening

For many women, especially in Asia, a lighter skin is important. In contrast to Europeans and Americans, they do not roast themselves on the beach – they protect their sensitive skin against sun-borne wrinkles.

It is the amino acid tyrosine, which protects the skin against UV irradiation, by producing the black-brownish pigment melanin. Melanin

synthesis starts after exposure to UV radiation. The enzyme tyrosinase and other enzymes catalyze the production of melanin. Biopsies, taken from volunteers in Düsseldorf, revealed that Pycnogenol® reduces the amount of RNA's encoding for tyrosinase and two other enzymes, needed for synthesis of melanin. (Ref. 5) So Pycnogenol® inhibits biosynthesis of melanin both by inactivation of free radicals as well as by inhibition of the melanin forming enzymes. Pycnogenol® is very effective in protecting skin from sunburn, so not as much melanin is needed.

Pycnogenol® was four times more potent than the anti-melanogenic agent kojic acid. (Ref. 6) These results explain the mechanism for Pycnogenol®'s ability to reduce the size and intensity of dark spots on the skin, the melasma. A clinical study with 30 women revealed a significant lightening of the spots within 30 days. (Ref. 7)

Pycnogenol® Protects Against Sunburn

It is common knowledge that the sun is not always our friend. It helps us get needed vitamin D, but the skin also can accumulate UV damage over the years, leading to wrinkles and/or cancer. Sunburn is an inflammation resulting from the free radicals produced by the effect of sunlight on fats in the skin. Pycnogenol® helps protect against free-radical damage. Studies have shown that the time of exposure required for sunburn to develop can be increased with Pycnogenol®, but Pycnogenol® should not be your only protection from the sun. The use of sunblock, wearing a hat, and an awareness of exposure time are also important. However, be sure to get optimal vitamin D.

Pycnogenol® be used as an external sunscreen as well as an internal sunscreen. Professor Rohdewald participated in such studies himself.

He marked different areas of his forearm, applied different strengths of a Pycnogenol® gel to these areas, and then exposed the forearm to UV radiation. Pycnogenol® protected the skin in a dose-related manner, meaning that higher concentrations were better than lower concentrations. Professor Rohdewald remembers, "The red marks from the unprotected areas could be seen on my arm as long as half a year, whereas the areas protected by Pycnogenol® were not visible after one week."

Sunshield from inside

Studies conducted in the U. S. in 2001 by Professors Lester Packer (Univ. California, Berkeley) and Ronald Watson (Univ. Arizona, Tucson) show that four weeks of Pycnogenol® increased the time it took for skin to redden in the sun. (Ref. 8)

Healthy American volunteers were exposed to UV to determine their individual minimum erythema dose (MED), this is the amount of UV radiation needed to produce a light reddening of the skin. Thereafter, they consumed one milligram of Pycnogenol® per kilogram of their body weight for a week and the MED was again determined. Now, the UV dose required to produce reddening was 60% higher.

During the following week the Pycnogenol® dose was increased to 1.7 mg per kg body weight. This required the MED to be 85% higher than at start. Hence, Pycnogenol® will reduce UV-induced damage following oral intake. Pycnogenol® protects from sun damage, able to act from outside as a gel or cream and to act from inside as a tablet or capsule.

Skin Cancer

An Australian researcher investigated whether Pycnogenol® could help in preventing UV damage and skin cancer. It was already known from experiments conducted in Hungary that a Pycnogenol® cream reduced skin damage of shaved rats when it was applied before UV exposure. (Ref. 9) As Pycnogenol® absorbs UV, it seems to act in these experiments as a sun-screen. In Australia, a more sensitive strain of rodents was used. (Ref. 10)

As one can imagine, hairless albino mice are very susceptible to UV radiation. These mice develop skin cancer in a relatively short time. In a study, these mice were treated after radiation with UV topically with gels containing Pycnogenol®. As one could expect from the Hungarian experiments, the Pycnogenol®-containing gel brought relief from sunburn for the mice. The more concentrated the gel, the less was the inflammation caused by the UV. The radiation also damages immune competent cells in the skin. This UV-induced immunosuppression was reduced by the Pycnogenol® gels.

However, the most important result of this study was the protection against skin cancer. The daily application of a Pycnogenol® cream reduced the appearance of skin tumors to 85% relative to untreated, but also radiated, controls. Some mice remained tumor-free despite being UV-irradiated for 30 weeks. The untreated controls developed an average 5.2 tumors, whereas the treated mice averaged 3.5 tumors. The reduction of skin cancer is related to the protection of the immune system in the skin cells. Thus, Pycnogenol® acted in these experiments reducing both the symptoms of sunburn and the risk of skin cancer.

Oral Pycnogenol® protects against skin cancer

A similar investigation with the same strain of hairless mice, conducted in Greece, tested whether Pycnogenol®, given orally, could prevent the consequences of UV radiation. (Ref. 11)

The mice received Pycnogenol® daily in their drinking water, starting two weeks before UV radiation. The outcome of the study was very convincing: The Pycnogenol®-treated male mice developed skin tumors six weeks later than an untreated control group. Sixty percent of female mice in the control group developed tumors, but none of the mice in the Pycnogenol® group. Similarly, the number of tumors in male untreated mice was three, with Pycnogenol®, one. For the female mice, the picture was even clearer. There were two tumors in the untreated group versus simply none in those supplemented with Pycnogenol®.

These experiments impressively demonstrate that Pycnogenol® prevents skin cancer not only when applied as a gel, but also when given orally. This is quite understandable because the metabolites of Pycnogenol®, formed in the intestinal tract, are potent scavengers of free radicals and inhibitors of inflammatory substances. They travel with the blood stream, reach the skin and inactivate the free radicals originated from UV radiation.

A by-product of the investigation was the extension of lifespan of the treated mice relative to the control group. Pycnogenol® extended the lives of male mice by 22% and of female mice by 17%.

The results demonstrate a long-term protective effect of Pycnogenol® against skin carcinogenesis.

Scars and Ulcers

The main components of Pycnogenol®, the procyanidins, possess a high affinity to skin. For example, the astringent taste of powdered Pycnogenol® is evoked by the binding of Pycnogenol® to the surface of the tongue. One can easily imagine that this astringent could contract the skin at the edges of a wound, so that closure of the wound is accelerated.

As we have mentioned in earlier chapters, the active components of Pycnogenol® differ when ingested orally or absorbed through the skin. The absorption of Pycnogenol® constituents through the skin involves only the smaller molecules. The large procyanidins molecules cannot travel through the skin. However, these large molecules are tightly bound to the proteins of the skin and stick on the surface. They constrict the tissue and produce a bacteriostatic action, which contributes to wound healing.

It was shown by studies conducted in Hungary, that indeed the healing of wounds proceeds faster, when the wounds were treated with a gel containing Pycnogenol®. (Ref. 12) Wounds treated with the same gel only without Pycnogenol® were healed within 15 days. Identical wounds treated with Pycnogenol® in the gel at a concentration of five percent healed in only 12 days. Important, the diameter of scars was halved from four mm to less than two mm. This result could be of interest for cosmetic surgery.

Healing ulcers of the lower legs with Pycnogenol®

Whereas "normal," non-infected wounds of healthy persons as a rule present no serious problem, the situation is different for patients with poor circulation of the lower extremities. These patients may develop wounds on the lower legs, ulcers, as result of a combination of bad circulation, micro trauma, infections and chronic edema. These ulcers are difficult to treat, and healing is a lengthy process.

A clinic in San Valentino, Italy, with a great experience in treating venous diseases, was interested in testing Pycnogenol® for the treatment of such ulcers. The idea was that treatment of the underlying disease, the venous problems, would probably help to cure the ulcers.

Remarkably, the venous ulcers actually contracted faster following the intake of Pycnogenol® capsules. After 6 weeks, the ulcer sizes in the Pycnogenol® group was about one-third of the ulcer sizes in the control group. (Ref. 13)

At first glance it is astonishing that these skin ulcers responded to Pyc-nogenol® which was not applied externally to the ulcers. The Pycnoge-nol® had to act from inside. One reason is the improvement of venous circulation with Pycnogenol®. We demonstrated this in several clinical investigations, as we show in Chapter 10. Better circulation means bet-ter nutrition and oxygenation of the tissue helping the growth of new cells. Furthermore, the constituents of Pycnogenol® have several anti-in-flammatory actions. They scavenge the free radicals, produced from the inflamed wound. They block the formation of a range of inflamma-tory mediators and inhibit inflammatory enzymes. Hence the chronic inflammation of the wound can be reduced. Finally, Pycnogenol® inhib-its enzymes, collagenases, which destroy collagen, an important building block of our tissue. Collagenases are produced in large quantities during inflammation.

The Italian researcher was not satisfied with just the oral treatment of ulcers. He asked himself – learning from the successful Hungarian treat-ment of wounds with a gel – whether an external treatment with Pycnoge-nol® could accelerate the healing of ulcers even more.

Prof. Belcaro simply opened the Pycnogenol® capsules and applied the content in a fine layer on the wound. In addition, patients had to take the capsules orally. This combination of inside plus outside treatment was very successful and the ulcer sizes decreased more rapidly.

After six weeks, the ulcer was cured: Area size zero!

The ultimate challenge: treatment of diabetic ulcers

Diabetic ulcer, mostly foot ulcer, is a very serious problem for patient and doctor. Approximately 25% of all diabetic hospital submissions in USA or UK are caused by ulcers. The risk for amputation of lower extremities is 15 to 46% higher for diabetic patients.

There were good chances for a success of the approach with Pycnogenol®. First of all, it was very effective in healing the venous ulcers. And as in the case of diabetic ulcer, the anti-diabetic action of Pycnogenol® is going to the roots of the disease, the diabetes.

In the Italian clinic a comparison was made with diabetic patients whether Pycnogenol® capsules given orally, or applied locally as a powder or both treatments at the same time were the most promising way to treat the diabetic ulcer. (Ref. 3)

As expected, the healing of the diabetic ulcers was more difficult compared to the venous ulcers; a complete healing could not be achieved. However, the intake of capsules reduced ulcer area by more than 10% compared to the control group. The content of the capsules, applied locally, produced another 10% more. The combination of internal and external treatment caused a decrease in ulcer size of 68%.

Such an effect obtained in just 6 weeks looks phenomenal and recommends the development of a powder or foam formulation for treatment of wounds and ulcers.

Pycnogenol® is more than an oral cosmetic, it can be a limb-saver as well.

Chapter References

1. Grimm T, Schäfer A, Högger P. Antioxidant activity and inhibition of matrix metalloproteinases by metabolites of maritime pine bark extract (Pycnogenol®). J Free Radic Biol Med 36: 811–822, 2004

2. Marini A, Grether-Beck S, Jaenicke T, et al. Pycnogenol® Effects on Skin Elasticity and Hydration Coincide with Increased Gene Expressions of Collagen Type I and Hyaluronic Acid Synthase in Women. Skin Pharmacol Physiol 25: 86–92, 2012

3. Belcaro G, Cesarone MR, Errichi BM, et al. Diabetic Ulcers: Microcirculatory improvement and faster healing with Pycnogenol®. Clin Appl Thromb Hemost 12: 205–212, 2006

4. Segger D, Schönlau F. Supplementation with Evelle® improves smoothness and elasticity in a double blind, placebo-controlled study with 62 women. J Dermatolog Treat 15: 222–226, 2004

5.   Krutmann et al – in preparation.

6.   Kim YJ, Kang KS, Yokozawa T. The anti-melanogenic effect of Pycnogenol® by its anti-oxidative actions. Food and Chemical Toxicol 46: 2466–2471, 2008

7.   Ni Z, Mu Y, Gulati O. Treatment of melasma with Pycnogenol®. Phytother Res 16: 567–571, 2002

8.   Rhin B, Saliou C, Bottin MC, et al. From ancient remedies to modern therapeutics: Pine bark uses in skin disorders revisited. Phytother Res 15: 76–78, 2001

9.   Blazsó G, Gábor M, Rohdewald P. Antiinflammatory activities of procyanidin-containing extracts from Pinus pinaster Ait. after oral and cutaneous application. Pharmazie 52: 380–382, 1997

10.   Sime S, Reeve VE. Protection from inflammation, immunosuppression and carcinogenesis induced by UV radiation in mice by topical Pycnogenol®. Photochem Photobiol 79: 193–198, 2004

11.   Kyriazi M, Yova D, Rallis M, et al. Cancer chemopreventive effects of Pinus maritime bark extract on ultraviolet radiation and ultraviolet radiation-7,12 dimethylbenz(a) anthracene induced skin carcinogenensis of hairless mice. Cancer Lett 237: 234–241, 2006

12.   Blazso G, Gabor M, Schönlau F, Rohdewald P. Pycnogenol® accelerates wound healing and reduces scar formation. Phytother Res 18: 579–581, 2004

13.   Belcaro G, Cesarone MR, Errichi BM, et al. Venous Ulcers: Microcirculatory Improvement and Faster Healing with Local Use of Pycnogenol®. Angiology 56: 699–705, 2005

Chapter Eight | Living Better Longer: Additional Anti-Aging Effects of Pycnogenol®: Memory Enhancement and Longevity

We have discussed how Pycnogenol® reduces our risk of age-related diseases such as heart disease, cancer and arthritis, and we have examined how Pycnogenol® keeps our skin younger. There are other ways that Pycnogenol® helps protect our bodies against the ravages of time including maintaining brain function and healthy longevity.

After one of Professor Rohdewald's lectures in the People Republic of China he was asked: "What can I expect from taking Pycnogenol®? He tried to make a long story short and answered through his interpreter, "If you take Pycnogenol® on a regular basis, you will live longer in better shape." Coincidently, this was essentially the title of one of Dr. Passwater's books on Pycnogenol®, "Live Better Longer." Professor Rohdewald went on to explain why Pycnogenol® could help people live a longer and healthier life, but the questioner did not seem to be interested in the "technical" explanations about the impact of free radical scavenging and anti-inflammatory activity on life expectancy.

Some weeks later, Professor Rohdewald received a message from China: "The study to demonstrate an extended life span through Pycnogenol®'s power is initiated! And a question: Which dose of Pycnogenol® is recommended for fruit flies?"

The answer to this question is not that easy: How to transfer the dosage of 1–2 mg per kg body weight for humans as the highest developed species down to the body weight of less than 1 mg of a fruit fly? However, there is an accepted formula to transfer dosages from one species to another and the study soon was underway in China.

Messages We Can Use Even from Studies with Fruit Flies

Of course, humans are not very comparable with fruit flies, but, surprisingly to many, humans have a lot of genes in common with fruit flies. Therefore, a positive outcome of the Chinese fruit fly study could possibly provide some information for humans.

Scientists often use fruit flies in studies because they have a short, but energetic, life time. They are an excellent model to study the impact of environmental and genetic factors on their lifespan.

Some lifespan studies of fruit flies provide exemplary insights

One male fruit fly living in a cage together with several females has quite a short lifespan. One male fruit fly living together with just one female fruit fly lived significantly longer. An isolated male fruit fly has the most extended life span.

We draw no conclusions from this

In the Chinese study, the male and female fruit flies were investigated separately. The female fruit flies live longer than the male, which is comparable to human observations that, on average, men have shorter lifespans than women. However, feeding male fruit flies with Pycnogenol® prolonged their mean lifespan by 13%. Transferred to humans this would correspond to an increase of 9 years of life. The lifespan extension for the female fruit flies by feeding Pycnogenol® was only 6%, which would mean an increase of about 4 years, if we transfer the results to humans. (Ref. 1)

So the male fruit flies profited more from Pycnogenol® than the females, but, the male flies could not catch up with the females having still a longer survival time, even after Pycnogenol®'s better support for the males.

These results were convincing enough for the Chinese Ministry of Health in 1999 to allow the following health claim for Pycnogenol®: "Pycnogenol® is suitable for people who want to slow down their aging process."

The decision to permit such a strong claim takes into consideration the influence of free radicals on aging, which could be demonstrated in several investigations not only in fruit flies, but also in other species. The decision of the Chinese Ministry of Health to accept basic scientific findings from a fast-reacting biological system instead of requiring extremely costly, long-term studies with elderly human populations is well warranted. Such human studies are impractical and will never be done. The ability of Pycnogenol® to scavenge all types of free radicals and to increase the production of anti-oxidative and anti-inflammatory enzymes provides an explanation for this extension of mean life span.

The fruit fly experiment provides some verification of the first part of Professor Rohdewald's quick response, "You will live longer in a better shape," to the question posed by an audience member after his lecture in China. Now let's examine whether Pycnogenol® can help us age better.

One important point for aging in good shape is the function of our brain under white hairs. Is it possible to slow aging of brain with Pycnogenol®? Furthermore, is it possible that Pycnogenol® can even help the memory of young students? Tests show that the answer is a definite and significant "Yes."

Studies on memory

The first answer was given by old mice. A comparison was made between normal and senile mice. Groups of mice had to make simple tests of their memory and learning. Youngsters were tested against the oldies – and against oldies fed with 2 different doses of Pycnogenol®.

As expected, old mice forgot everything after a short period of learning, after 6 days they forgot everything. But, reinforced with Pycnogenol®, the old mice mastered their lessons, better and better, the more Pycnogenol® they consumed. 60% of the old mice could master the tests over a period of 10 days after intake of 10 mg/kg body weight; with 5 mg/kg, still 50% of the old mice could remember the tests while the rate of the successful young mice was 70%. (Ref. 2)

The improvement in memory and learning of the senile mice was very convincing, hence we jumped to human beings.

A better memory has been found also in tests with volunteers in Italy.

Healthy professionals in the age between 35 and 55 years and students between 18 and 27 years were tested in groups of 33 to 55 participants.

The cognitive function of the participants was evaluated in a range of tests, reflecting abilities for daily life.

The professionals could significantly improve memory as well as professional daily tasks after three months with 150 mg Pycnogenol®.

Also parameters for mood, as alertness, anxiety and contentedness were ameliorated in contrast to the control group. (Ref. 3)

Students, consuming 100 mg Pycnogenol® over a period of 8 weeks had a significantly higher success rate in exams at the university: in the control group 11% failed, in the other group only 6%. (Ref. 4)

Thus, Pycnogenol® improved the memory and cognitive function also in men.

Finally, also in our investigations with women in menopause, an improvement of memory was reported. (Ref. 5)

These results support altogether the suggestion that Pycnogenol® improves brain function for healthy young people as well as middle-aged men and women. This represents a great contribution to the goal: Living in better shape. If we add the ability of Pycnogenol® to improve also physical fitness as we discuss in Chapter 14 on Top Performance, we may propose the use of Pycnogenol® for body and soul.

Better results in the US Army physical fitness test as well as in triathlon using Pycnogenol® (Ref. 6) suggest that the elderlies could improve their fitness, too.

Now let's look at how Pycnogenol® relieves the symptoms of allergies.

Chapter References

1. Shuguang L, Xinwen Z, Sihong X, Gulati OP. Role of Pycnogenol® in aging by increasing the Drosophila's life-span. Eur Bull Drug Res 11: 39–45, 2003

2. Liu F, Zhang Y, Lau BHS. Pycnogenol® improves learning impairment and memory deficit in senescence-accelerated mice. J Anti Aging Med 2: 349–355, 1999

3. Belcaro G, Luzzi R, Dugall M, et al. Pycnogenol® improves cognitive function, attention, mental performance and specific professional skills in healthy professionals age 35 – 55. J Neurosurg Sci 58 pages to come 2014

4. Luzzi R, Belcaro G, Zulli C, et al. Pycnogenol® supplementation improves cognitive function, attention and mental performance in students. Panminerva Med 53: 75–82, 20115.

5. Yang HM, Liao MF, Zhu SY, et al. A randomized, double-blind, placebo-controlled trial on the effect of Pycnogenol® on the climacteric syndrome in perimenopausal women. Acta Obstet Gynecol Scand 86: 978–985, 2007

6. Vinciguerra G, Belcaro G, Bonanni E, et al. Evaluation of the effects of supplementation with Pycnogenol® on fitness in normal subjects with the Army Physical Fitness Test and in performances of athletes in the 100-minute triathlon. J Sports Med Phys Fitness 53: 644–654, 2013

Chapter Nine | Allergies, Asthma and COPD

Pycnogenol® provides health benefits other than those the nutritional support of age-related conditions as well. The early popularity of Pycnogenol® was due to its ability to reduce the symptoms of allergies. Decades before Pycnogenol®'s antioxidant and anti-inflammatory roles were known, it was used successfully in Europe to control hay fever and other allergies.

Allergies are hypersensitive reactions that occur when the body comes in contact with harmless substances that the body perceives as harmful. Substances that cause these reactions are termed allergens. When a hypersensitive person comes into contact with an allergen, the body releases histamine in an attempt to fight off the allergen. This release of histamine triggers the symptoms so common to allergies – inflammation, sneezing, runny nose, and itchy eyes.

Pycnogenol® directly blocks histamine release. Pycnogenol® increases the uptake and re-uptake of histamine into its storage granules in the mast cells of the immune system, where it's out of the way and can't cause misery. (Ref. 1)

Generally, antihistamines work differently by interfering with the attachment of histamine to cells after it has been released. It's more efficient to prevent histamine release in the first place than to try to keep released histamine away from its receptors on target cells. Pycnogenol® is effective against allergies without producing side effects as drowsiness and dry mucous membranes.

When Professor Rohdewald was first asked to evaluate Pycnogenol® by his German physician friend, the clinical studies at that time centered around allergy symptom reduction and venous insufficiency. This is an example of a nutrient relieving symptoms in mild allergies or being an adjunct therapy along with medication for acute cases. Years later, Pro-

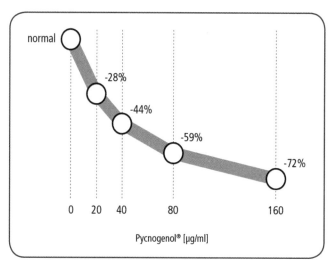

Figure 9.1. Pycnogenol® inhibits histamine release from mast cells. (Modified from Ref. 1)

fessor Rohdewald wanted to re-visit the studies using the latest scientific methods.

Block of Histamine Release in Cell Culture

In vitro studies. Professor Sharma from the Department of Pharmacology of the University of Dublin showed that Pycnogenol® inhibited histamine release from mast cells. The inhibition was more pronounced than with the antihistamine drug chromoglycate. (Ref. 1)

Korean researchers corroborated these cell culture findings regarding histamine release with laboratory animal studies. Feeding of rats with Pycnogenol® inhibited an allergic reaction of the skin. The scientists investigated the mechanism of the blocking of histamine release more thoroughly and concluded that Pycnogenol® was effective in reducing swelling and skin reddening in mast cell-mediated allergic diseases. (Ref. 2)

Clinical Trials

It is not that easy to conduct an appropriate clinical trial investigating how Pycnogenol® provides nutritional support that relieves allergic symptoms. As Professor Rohdewald recollects, "We started a costly study in Canada, fulfilling all the requirements of good clinical practice. (Ref. 3) The purpose of the study was to substantiate the anecdotal reports and the antihistaminic actions from the laboratory. Would Pycnogenol® suppress symptoms of hay fever during the birch pollen season in a placebo-controlled study? For this study, allergic volunteers had to take Pycnogenol® or placebo some weeks before predicted start of the birch pollen season. Unfortunately, recruitment of volunteers was insufficient. The small numbers of patients (20) in the Pycnogenol® and placebo groups produced statistics that only allowed the description of trends."

"One could see that patients got a better relief from symptoms, when intake of Pycnogenol® started early before the pollen were flying. The eye symptom score was lower (0.13) in the Pycnogenol® group compared to the placebo group (0.2), corresponding to the lower level of specific immunoglobulin in the Pycnogenol® group (19 vs. 32)." (Ref. 3)

Such results suggested an additional benefit for those patients with allergic asthma, which affects nose and eyes besides the lung. Pycnogenol® will often help to alleviate symptoms of hay fever, when taken early as prevention. But, as in asthma, Pycnogenol® is not able to counteract the acute allergic attack.

Pycnogenol® in Asthma

Professor Rohdewald was very familiar with that disease. First of all, he occasionally suffered from asthma. Secondly, he was investigating drugs against asthma for a couple of years. From his experience, he knew that there were excellent anti-asthmatic drugs, namely corticosteroids for inhalation. But, he also knew that many people fear these drugs because of unwanted effects. There is nearly no reason to fear because unwanted effects from corticosteroid inhalation are extremely rare. However,

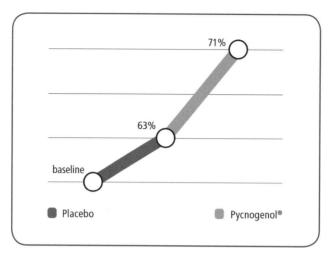

Figure 9.2 Pycnogenol® allows easier breathing (FEV1). (Ref. 4)

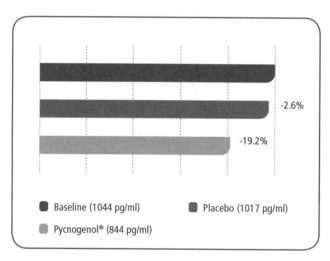

Figure 9.3. Pycnogenol® lowers blood leukotriene (inflammation) levels. (Ref. 4)

some trust on principle only natural substances and don't like powerful synthetic drugs. This group of people tends to withhold everything they perceive as being unhealthy from their children. This group would be happy to have an alternative option on a botanical basis to prevent and to treat asthma.

Asthma is a chronic inflammatory disease of the lung. It had been shown in a plethora of test systems, that Pycnogenol® is a broadly-acting anti-inflammatory agent. So it should be helpful also in asthma. And as it blocks the release of histamine, it should help in case of allergen-induced asthma.

Nibbling grandma's mouth spray improves asthma

A report came from Finland from a Finnish manufacturer that produced a Pycnogenol®-containing mouth spray. This spray attenuates inflammation of the gum and was used by a grandma to get her artificial teeth fixed in the morning. The spray contained glycerol as solvent for Pycnogenol®. The two grandchildren, imitating the daily routine of grandma, realized that the spray tastes sweet – from glycerol. So they were nibbling all day from Grandma's spray. The grandchildren had asthma and Grandma reported to the pharmacist that the spray improved the symptoms of asthma. This was one more reason to start a clinical study to get solid information whether Pycnogenol® helps patients with asthma.

The Iranian-American cooperation

A clinical study was started as a result of an unusual cooperation. Medical doctors in Iran performed a clinical study with asthmatic patients. The study was designed and evaluated in the USA. Twenty-two asthmatic patients received either Pycnogenol® 1 mg/lb/day or a placebo. Neither the physicians nor the patients knew who received the placebo or the Pycnogenol®. In the placebo group, neither asthma symptoms nor lung functions were changed while the Pycnogenol® group improved in both respects. (Ref. 4)

The study used the "cross-over" procedure to help eliminate individual variances. In a crossover, the groups stop taking all pills for period of time to allow any active substances being tested to wash out. Then they are given the opposite pills. Those who were formerly receiving the placebo were then unknowing given Pycnogenol® ... and vice versa.

The patients' airway function was assessed by the "forced expiration volume in 1 second"(FEV1), by means of a spirometer. The subject fills his lungs and the air volume exhaled fast within 1 second is measured. The exhaled volume is expressed relative to the total lung volume, so the FEV1 value represents the percentage of a patient's lung volume he can exhale in a second. Naturally, the percentage is lower in asthmatics as their airways are constricted, and breathing is aggravated. After 4 weeks treatment with Pycnogenol® the patients could exhale 71% of their lung volume as compared to 59% at the start of the study and 63% in response to placebo, respectively.

The severity of asthma symptoms was rated on a four-point scale, ranging from symptom free (0) in several steps to mild intermittent (1), to a moderate intermittent form (2) up to a severe persistent form (3). Symptom scores were in average 2.23 before treatment and 2.79 while receiving placebos, which are rated as being a "severe persistent" form. In patients taking Pycnogenol® the average symptom severity score was significantly lowered to 1.75, a "moderate persistent" form.

The improvement of airway function was paralleled by a reduction of leukotrienes, pro-inflammatory mediators, in the blood. Leukotrienes attract immune cells to the bronchi and activate them. This causes bronchial constriction and airway obstruction in asthma.

Pycnogenol® significantly reduced the leukotriene values in the blood of patients, as compared to both baseline values as well as placebo medication. As expected, the placebo had no significant influence on leukotriene levels in the blood. Pycnogenol® was well tolerated, only one patient experienced gastrointestinal discomfort; however, this occurred only during the first 3–4 days. The patients generally noted an improvement of their breathing ability when they received Pycnogenol®.

The same result was obtained when the patients of the Pycnogenol® group had to switch to placebo and vice versa. The Pycnogenol® group did significantly better. This sophisticated placebo-controlled, double-blind

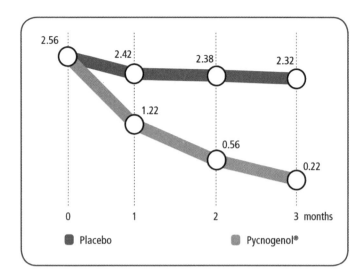

Figure 9.5. Lowered requirement for rescue puffs per 24 hours. (Ref. 5)

cross-over study gave us a strong hint that Pycnogenol® improves asthma symptoms. However, the group sizes were small. Despite the fact that the probability of a false positive result was less than one per thousand, every scientist will call for more patients to be studied.

Less asthma symptoms in children

Hence we made a second study with children having mild-to-moderate asthma in the USA. The vast majority of asthmatics have developed the disease already during childhood, most of them before they reached the age of five years. In many cases children develop hay-fever, which thereafter progresses to asthma.

Asthma medication with children is a sensitive issue, representing a challenge for everybody involved; the treating physician, the parents, and the children themselves. This results from the high symptom variability in children, further complicated by the dynamics of the development and growth of the child. Parents feel uncomfortable with having their child permanently take prescribed medications.

The clinic had received positive feedback from asthmatic patients using Pycnogenol®. The 60 young patients in this double-blind, placebo-controlled study confirmed the results of the study with the adults. (Ref. 5) The use of Pycnogenol® 1 mg/lb/day over three months steadily improved lung function; consequently, their use of rescue inhalers dropped down to nearly zero.

A double-blind, placebo-controlled study has investigated 60 children with mild to moderate asthma, aged 6 to 18 years, over a period 3 months. (Ref. 5) A minority of nine patients took oral medication with Accolate® (Zafirlukast). All patients were depending on rescue inhalers (with albuterol) to control occurring asthma attacks. Thirty children were assigned to treatment with Pycnogenol® (1mg/lb/day) and another 30 children to the control group receiving placebo for three months. One month prior to treatment was taken as a run-in period to establish baseline conditions.

Triggers for asthmatic attacks

- ☐ Pollen
- ☐ Perfume
- ☐ Smog
- ☐ Stress
- ☐ Animal Hairs
- ☐ Cold Air
- ☐ Strenuous Exercise
- ☐ Emotional stress
- ☐ Dehydration
- ☐ Inflammation
- ☐ Histamine (in wine, cheese)
- ☐ Food allergies
- ☐ Ozone

The study showed that ease of breathing improved significantly already after 1 month treatment with Pycnogenol® supplementation, as measured using the FEV1 method. Breathing was expressed as percentage of total lung volume that can be exhaled in a second. Breathing ability was further improving after two months and three months treatment, whereas placebo had no effect at any time.

The severity of asthma symptoms was rated on a four-point scale. At baseline, mean symptom scores was 2.3 which is between 2 = moderate ("somewhat disturbing") and 3 = severe ("interfering with daily activities"). Symptoms gradually decreased during Pycnogenol® treatment and reached 0.2 by the end of the study, the children thus being almost symptom free. In contrast, the placebo-treated group had symptoms only marginally improved that remained above 2 until after the completion of the study.

The improvement of airway function was paralleled by a reduction of inflammatory mediators (leukotrienes), tested from the urine of the patients. The leukotrienes cause the inflammatory condition and bronchial constriction. Pycnogenol® significantly reduced leukotriene values already after one month and further decreased them throughout the study period. As expected, treatment with the placebo had no influence on leukotriene levels. The most compelling outcome of the study was the dramatically reduced necessity of using rescue inhalers as severe asthma attacks appeared much less frequently. After one month, 8 out of 30 children taking Pycnogenol® didn't require rescue inhalers at all anymore, and the number increased to 12 and 18 completely off the inhaler after two and three months treatment, respectively. The researcher concluded that "Pycnogenol® is an effective and safe nutritional approach for children to manage mild to moderate asthma."

The positive effects of Pycnogenol® were highly significant. As in the foregoing study, patients did not report any adverse effects. The children tolerated the Pycnogenol® very well.

The ultimate challenge: Is Pycnogenol® as an additive to best possible care able to produce an even better relief of symptoms

A group of doctors from the University of Siena/ Pescara got the message from the effects of Pycnogenol® on asthma and challenged the power of Pycnogenol® in a controlled study. (Ref. 6) They were keen to see whether the current best possible treatment of asthma could be improved by including Pycnogenol® as an adjunct. In this study, 65 patients with mild to moderate asthma got treatment with a very potent, inhaled corticosteroid (ICS). Half of the patients also received Pycnogenol® in addition to the ICS. After 6 months, there was a slight, but significant positive effect of Pycnogenol®. There was a reduction of about 50% in diverse symptoms of asthma including night-time awakening, use of bronchodilator, and days having asthma attacks. Important, Pycnogenol® improved the quality of life.

The authors concluded that the signs and symptoms of asthma are easier to control with Pycnogenol®.

A Personal Remark by Professor Rohdewald

Just as a side note, Professor Rohdewald concurs with these studies from a personal point of view. "Yes, I agree with this conclusion. I have regularly taken Pycnogenol® for more than ten years and now I don't suffer from real asthma attacks anymore. I seldom use bronchodilators and I need ICS fewer than ten days in a year. So I can recommend trying Pycnogenol® in case of mild to moderate asthma for continuous protection, not as a rescue medication. Pycnogenol® can only **prevent** the asthma symptoms, as it prevents the histamine release and blocks the inflammatory mediators. But it is not able **to reverse** the bronchoconstriction during an asthmatic attack."

Pycnogenol® offers an alternative choice for patients with mild asthma or beginning asthma, especially for those who prefer to use botanicals instead of synthetic drugs.

Chronic Obstructive Pulmonary Disorder (COPD)

Also known as smokers cough, this chronic obstructive lung disease is a heavy burden. 210 million people are currently known to live with COPD. But estimates are that 50% of cases of COPD are undetected and 25% are misdiagnosed, so that real figures are considerably higher.

Five-to-thirteen percent of Europeans and seven percent in the USA suffer from chronic cough. Overall costs caused by the COPD amount in the USA to 32 billion dollars annually. COPD is at presence the fifth leading cause of death worldwide; it is expected to become the third leading cause of death by 2020.

These are good reasons to outline how Pycnogenol® could help the victims of COPD.

Mechanism and symptoms of COPD

COPD patients suffer from a chronic inflammation of the lung, from a progressive destruction of their lung tissue and a systemic inflammation. The disease gets started by oxidative stress, caused by smoking or other environmental factors. The oxidative stress attracts white blood cells. The white blood cells liberate pro-inflammatory substances and protease enzymes. The protease enzymes destroy lung tissue, resulting in emphysema, and ultimately, in death.

The great number of inflammatory substances leak from the lung to the whole organism and they are not sufficiently inactivated by the body's own defense systems. As a result, the muscles of the lung are attacked by free radicals and become weaker.

The COPD patient has difficulty breathing because of less active lung tissue, less power to breath, and less opening of the airways. Unfortunately, the destruction and remodeling of the airways is irreversible.

Pycnogenol® may fight COPD in several ways

Pycnogenol® reduces the oxidative stress, as we have shown in several clinical investigations. It doubles the amount of anti-oxidative enzymes inside our cells.

Just as important, Pycnogenol® blocks the production of a series of the most important pro-inflammatory molecules.

Last but not least, Pycnogenol® inhibits the dangerous protease enzymes, which are responsible for the destruction of lung tissue. Furthermore, because of its antibacterial and antiviral properties, Pycnogenol® can help prevent lung infections that frequently afflict COPD patients.

125

Therefore, one can expect that Pycnogenol®, added to bronchodilators, antihistamines and inhaled glucocorticoids, will help to slow down the progress of the disease.

However, COPD is mostly caused by smoking. In such circumstances, the progression of the disease is easily stopped by non-smoking.

In this chapter we have examined the nutritional support and adjunct therapy of Pycnogenol® in allergies, asthma and COPD. In the next chapter we will look at Vein Health and Circulation.

Chapter References

1.   Sharma SC, Sharma S, Gulati OP. Pycnogenol® as an adjunct in the management of childhood Asthma. Phytother Res 17: 66–69, 2003

2.   Choi YH, Yan GH. Pycnogenol® inhibits immunoglobulin E-mediated allergic response in mast cells. Phytother Res 23: 1691–1695, 2009

3.   Wilson D, Evans M, Guthrie N, et al. A randomized, double blind, placebo-controlled exploratory study to evaluate the potential of Pycnogenol® for improving allergic rhinitis symptoms. Phytother Res 24: 1115–1119, 2010

4.   Hosseini S, Pishnamazi S, Sadrzadeh MH, et al. Pycnogenol® in the management of asthma. J Med Food 4: 201–209, 2001

5.   Lau BHS, Riesen SK, Truong KP, et al. Pycnogenol® as an adjunct in the management of childhood asthma. J Asthma 41: 825–832, 2004

6.   Belcaro G, Luzzi R, Cesinaro Di Rocco P, et al. Pycnogenol® improvements in asthma management. Panminerva Med 53: 57–64, 2011

Chapter Ten | Vein Health and Circulation

We have discussed cardiovascular circulation in Chapter Two on heart disease with emphasis on arterial health and circulation. Now let's look at the other half of the circulation, venous health and circulation. The veins in our body return blood back to the heart. Unlike arterial blood, which is actively transported by ejection from the heart, venous blood is passively transported by compression of veins through a long series of valves allowing movement only in one direction, leading to the heart. We have also mentioned several times that the early uses of Pycnogenol® centered on venous insufficiency and allergies in which the results are rapid and easily discerned. To date, more than 25 clinical studies on over 1,000 people have verified Pycnogenol®'s health benefits related to venous health. (Ref. 1)

Venous Insufficiency and Swollen Ankles

Swollen ankles may result from many diseases. For the elderly, the most frequent cause for edema of the lower limbs is the venous insufficiency. If the circulation in the venous system is impaired, blood is not sufficiently pumped upwards. The extra weight of this accumulated blood increases the blood pressure in the lower limbs. Blood in veins of the lower legs have the longest distance to travel and experiences the greatest gravity counterforce. When vein valves do not totally withstand the pull of gravity, blood will pool in the veins of the lower limbs. This can happen while sitting in an upright position or standing in one place for a long period of time. The venous blood, not actively pumped by the movement of the leg muscles, accumulates in the lower limbs. The pressure inside the veins damages the epithelial wall, liquid is squeezed out from the bloodstream (without the red blood cells) into the surrounding tissues. The surplus of liquid inside the tissue causes swelling, and edema develops. In chronic venous insufficiency, the permanent high venous pressure causes oxidative stress, damage of blood capillaries and finally a chronic inflammation of the veins. This inflammation may lead to destruction of tissue, restricted microcirculation and thrombosis.

In addition to swelling, typical symptoms of venous insufficiency can include itching skin of the legs and feet, aching, cramping or tiredness of legs, skin discoloration and appearance of new varicose veins. If left untreated, the situation may further deteriorate with tiny capillaries in the skin bursting, leading to a brownish discoloration and badly healing wounds (ulcers). A serious risk is the possibility of suffering a thrombosis (blood clot), which may clog the vein or even travel to other parts of the body leading to great harm.

In general, extended periods of standing or sitting will increase the risk for developing chronic venous insufficiency (CVI). Statistics suggest that women are more commonly affected, even more so during pregnancy. Being overweight or suffering a deep vein thrombosis may lead to developing CVI. Because faulty venous valves cannot be healed or surgically repaired, CVI needs to be dealt with as early as possible to halt the progression of the disease.

Pycnogenol® works in two independent ways to prevent and diminish fluid accumulation in tissue. Pycnogenol® strengthens capillary walls and makes them more resistant to pressure with consequently lowered release of unwanted fluids into tissues. It is important to note that blood plasma (blood without red blood cells) is required to pass through blood vessel walls to nourish organs with nutrients and supply oxygen.

As we have discussed in Chapter Two on Heart Disease, Pycnogenol® also improves endothelial function with enhanced nitric oxide synthesis, which in turn reduces blood vessel constriction, which allows the blood to flow easier.

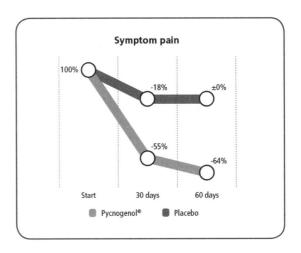

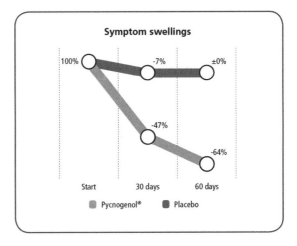

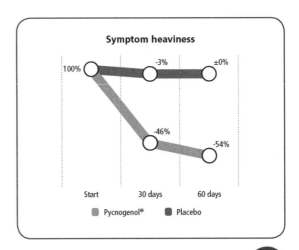

*Figure 10.1 Pycnogenol®
significantly improves
the symptoms of edema
of the ankles (ankle
swelling). (Ref. 2)*

Many clinical studies confirm that Pycnogenol® alleviates chronic venous insufficiency

When Professor Rohdewald first evaluated the clinical studies on Pycnogenol® at the request of his physician friend, there were several published in French. Since 1989, 17 additional studies with 1,038 patients have investigated the role of Pycnogenol® in the treatment of chronic venous insufficiency. All these clinical studies registered a significant reduction of edema as well as the feeling of heaviness and pain of the legs, compared to control or to placebo. The decrease of ankle and foot edema was objectively measured either by measuring circumference of the ankles or by a water displacement technique.

In one typical double-blind, placebo-controlled study Pycnogenol® significantly improved three of the most common symptoms related to the swellings, the perception of leg heaviness and the resulting leg pain already after 30 days treatment. (Ref. 2) Taking Pycnogenol® for an additional 30 days provided further symptom relief. Placebo contributed only marginally to improvement of symptoms. The results on individual clinical symptoms, pain, swellings and the perception of leg heaviness for both groups are illustrated on the facing page.

Pycnogenol® outperforms drugs

In a clinical comparison of Pycnogenol® to a horse chestnut extract (Venostasin), Pycnogenol® was found to more effectively reduce ankle swelling and significantly improved subjective symptoms better. (Ref. 3) In another study, Pycnogenol® was compared with a combination of the flavonoids diosmin and hesperidin (Daflon). In this study, Pycnogenol® produced a faster and better relief of symptoms and improved microcirculation, resulting in a significantly better oxygenation of the tissue. (Ref. 4) In 86 venous insufficiency patients, patients were given either a daily regimen of 1 g Daflon® or 150 mg Pycnogenol® over a period of eight weeks. The ankle swelling, as judged by strain gauge plethysmography, was found to be significantly lowered by 24% with Pycnogenol® already after four weeks treatment, which was not the case with Daflon®. After

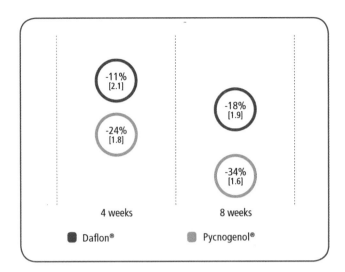

Figure 10.2. Ankle swelling reduction with Pycnogenol® compared to a pharmaceutical.

eight weeks treatment Pycnogenol® was found to be significantly more effective reducing edema than Daflon®.

This study further investigated the patient's symptoms of pain, restless legs, skin alterations and subjective feelings.

Pycnogenol® was further shown to be helpful for significant improvement of symptoms in individuals presenting with more severe venous insufficiency. (Ref. 4) Such cases are characterized by ambulatory venous pressure larger than 50 mmHg. In a controlled study Pycnogenol® proved to be helpful for all investigated symptoms, such as the sensation of leg restlessness, pain, edema and skin discoloration.

After two weeks of Pycnogenol®, the symptoms improved repidly, with a noticeable 42% reduction. After eight weeks Pycnogenol® significantly improved even severe cases of venous insufficiency.

In a third study, Pycnogenol® was added to Troxerutin, a semi-synthetic flavonoid. The combination of 20 mg Pycnogenol® and 470 mg Troxerutin was found to be superior to 600 mg pure Troxerutin in ameliorating symptoms. (Ref. 5)

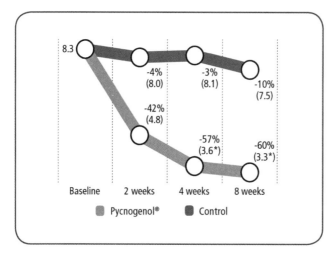

Figure 10.3. Symptom score [0–10] covering edema, sensation of limb "restlessness," pain, swellings and skin discoloration (Ref. 4)

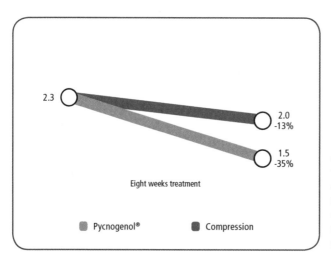

Figure 10.4. Comparison of rate of ankle swelling between compression stockings or Pycnogenol®. (Ref. 8)

Pycnogenol® Improves Venous Elasticity

Thirty subjects with varicose veins and chronic venous insufficiency used 150 mg Pycnogenol® per day for a period of three months before surgery for vein stripping. Venous segments were tested after surgery for elasticity by stretching under defined conditions. The results were compared with veins obtained from patients with varicose veins, but without prior intake of Pycnogenol®. (Ref. 6) The measurement of stretching and the recovery after dilation indicated a significantly better recovery to the original size following the ingestion of Pycnogenol® compared to control. Thus, Pycnogenol® gives the vein walls a greater tonic recovery and elasticity allowing veins to recover after stress.

The ultimate test: a one year comparison of Pycnogenol® with elastic compression stockings

The most popular conventional method to treat chronic venous insufficiency is to wear compression stockings. The dense tissue of the stockings exerts a pressure against the venous pressure, thus mechanically preventing or reducing edema formation. To test whether Pycnogenol® is an option besides the compression stockings, a study was performed in Italy

Table 10.1 Time course of ankle swelling
156 patients with post-thrombotic venous insufficiency

	Compression Stockings	Pycnogenol® 150 mg	Pycnogenol® + Stockings 150 mg
	Ankle circumference (cm)		
Start	28	27.5	27
6 months	26.1	23.7	23
12 months	26.1	23.1	22.9

Table 10.2 Evolution of mean edema sign and symptoms score

	Compression Stockings	Pycnogenol® 150 mg	Pycnogenol® + Stockings 150 mg
Start	7.8	7.77	7.89
6 months	5.9	4.1	3.3
12 months	5.8	4.4	3.1

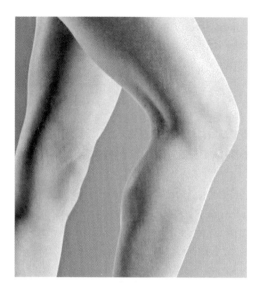

with three groups of patients who had at least once had a deep vein thrombosis: Group 1 had to wear elastic compression stockings, group 2 received 150 mg / day Pycnogenol®, and group 3 wore compression stockings as well as taking 150 mg Pycnogenol®. (Ref. 7)

The evaluations made after 6 months and at the end of the 12-month study revealed that the edema scores were significantly lower as compared to start in all groups. Really impressive was the significantly better reduction of edema with Pycnogenol® in comparison to compression stockings. The most effective treatment was the combination of stockings with Pycnogenol®. Please see Table 10. 1. Besides the reduction of edema, the microcirculation in the feet was improved, resulting in a better oxygenation of the tissue. None of the patients developed ulcers.

Limited tolerance of compression stockings

Wearing compression stockings is no pleasure, they are difficult to put on, especially in hot climate. Furthermore, the stockings are quite expensive because they have to be replaced after a few months.

Within the group wearing compression stockings 12 patients from 55 left the study. Two patients experienced a new deep vein thrombosis and had to quit the study, ten subjects refused to wear the compression stockings during the hot Italian summer. In the Pycnogenol® group, only two patients dropped out for non-medical reasons, in the "combined" group 5 patients dropped out. None of the patients in the groups 2 and 3 developed a new thrombosis.

Table 10.3 Pycnogenol® 150 mg /day versus compression stockings, second study

98 patients with chronic venous insufficiency

		Start	8 weeks
Stockings	Rate of Ankle swelling	2.3	2.0
Pycnogenol® (150 mg/day)		2.3	1.5
Stockings + Pycnogenol®		2.2	1.3
Stockings	Clinical severity score	8.4	5.7
Pycnogenol®		8.4	4.5
Stockings + Pycnogenol®		8.3	4.0
Stockings	Oxygen pressure	46.8	46.1
Pycnogenol®		46.2	50.1
Stockings + Pycnogenol®		47.1	51.1

Patients with severe venous problems got better symptom relief from Pycnogenol®

A second study by the same group used the same design as the previous study to treat patients without recent thrombosis, but with severe chronic venous insufficiency. (Ref. 8)

Patients in this study were also treated as three groups. Group one, wore compression stockings only. Group two received Pycnogenol® 150 mg / day only. Group 3 wore compression stockings and also took 150 mg Pycnogenol®. This study was for eight weeks. All the 98 patients completed the study. This study was conducted in the spring, making life wearing compression stockings easier. Also in this study of patients suffering from severe chronic venous insufficiency, Pycnogenol® had superior results over just wearing the compression stockings in decreasing ankle swelling and clinical symptoms. Pycnogenol® was found to produce a better oxygenation in the legs. Again, the combination of Pycnogenol® along with compression stockings produced the best results. The data are presented in Table 10.3.

Pycnogenol®, a well-tolerated aid for venous problems

The results of these two studies underline impressively the potency of Pycnogenol® for treatment of venous insufficiency and prevention of thrombosis. Pycnogenol® was very well tolerated in all these studies; none of the volunteers left a study because of any unwanted effects of the dietary supplement. The conclusion is that Pycnogenol® supplementation will improve quality of life for persons with problems in the venous circulation.

Traveling Long Distances – The Economy Class Syndrome and Jet Lag

Sitting, sitting and sitting for hours, in the middle of a row in the economy class is no pleasure at all. Some like to take their shoes off. Others avoid it because of the stress of getting their shoes on again. Swollen feet and swollen ankles are nearly normal at the end of a flight of six hours or more. Edema forms as result of restricted circulation. More important than uncomfortable swelling is the possibility to develop a deep vein thrombosis in the lower legs. This thrombosis is characterized by the common signs of inflammation: pain, swelling, heat and reddening. Besides these unpleasant symptoms, there is the danger that the thrombus becomes dislocated and travels as an embolus to the lung. A lung embolus is an immediate life threatening event. Consequently, it is very worthwhile to prevent thrombosis during long flights.

The question was whether Pycnogenol® would be able not only to reduce edema in the lower legs during flight, as has been demonstrated in many clinical studies, but whether Pycnogenol® could prevent deep vein thrombosis under the conditions of long flights. The idea was sim-

Table 10.4 Pycnogenol® Reduces Jet Lag Symptoms

	Controls Scores	Pycnogenol® Scores
Loss of appetite	7	3
Headaches	7	3
Fatigue	8	2
Disorientation	6	3
Upset stomach	6	5
Irregular sleep	9	3
Irritability	6	3
Irrationality	6	3
Bad performance	7	2
Alterations in well being	8	2
Duration of symptoms (hours)	39	18

ple – measurements before and after the flight. However, to do it, one had to test a large number of passengers in one continent before take off and on the other continent after landing using the same method of measurement.

The frequency of immobility-induced thrombosis is not that high, so many passengers had to be examined. The study was done with 198 passengers, one group received a placebo and another group received 200 mg Pycnogenol® three and six hours before the flight and one day later they received 100 mg of Pycnogenol®. (Ref. 9)

The passengers were tested for thrombosis before and after eight-hour flights. Taking into account the enormous effort for doing ultrasound investigation of the legs of each passenger at both sides of the Atlantic before and after flight, the result seemed meager at first sight.

In the Pycnogenol® group, just one case of non-thrombotic inflammation of a vein was observed. In the placebo group, four superficial thromboses and one deep venous thrombosis were found. However, if one takes into account that venous thrombosis may lead to lung embolism, it is a very good outcome: The risk for a potentially deathly event was zero in the Pycnogenol® group.

After the flight, the edema score for those passengers protected by Pycnogenol® increased by only 18% – compared to the high score of 58% in the control group. (Ref. 9)

Your permanent flight assistant: Pycnogenol®

It certainly is a good idea to increase the dose of Pycnogenol® before and during flights, as we do regularly. There's even another reason in addition to reducing swollen feet and the risk of thrombosis – jet lag. When crossing more than three or four time zones, the body may physically be in one time zone, but the body's biochemistry is lagging behind.

A number of symptoms may appear. First of all, disturbed sleep patterns – pretty alert in the middle of the night, but sleepwalking the day long. General well-being is decreased as well as mental performance. Some are easy to irritate, some have headache, others loss of appetite.

Within the framework of the above mentioned study, the researchers also noted less jet lag symptoms in the Pycnogenol® group.

Less jet lag with Pycnogenol®

To objectify these observations, a new investigation was started.

The 38 participating pilots, stewards and passengers had to take 150 mg Pycnogenol® for a week, starting 48 hours before take-off for a long distance flight. Flights were crossing meridians west-east or east-west. Scores for jet lag symptoms were evaluated within 48 hours after the flight. A control group of 30 untreated volunteers were evaluated for symptoms in the same way. (Ref. 10)

In the control group, the mean duration of jet lag symptoms was 39 hours, compared to just 12 hours in the Pycnogenol® group. Important, the severity of every symptom was significantly lower in the Pycnogenol® group. A summation of the study results is given in Table 10.4.

The study results suggested that persons prone to edema suffered more from jet lag. Therefore, it is hypothesized that some edema could also occur in the brain. Such a brain edema could be very small, not affecting clinical signs, but disturbing cognitive functions to some extent.

A deeper look into jet lag

To get a solid basis for this assumption, the study was repeated with the same design – 150mg Pycnogenol® for a week – but in addition a brain computer tomography (CT) was done within 28 hours after the flight. (Ref. 10)

Again, jet lag symptoms and ankle swelling were lower in the Pycnoge-nol® group compared to the untreated control group. Scans of the brain in the control group indeed showed the existence of minimal edema, covering less than 50% of the surface of the brain. In the Pycnogenol® group, less areas with edema were detected. A limited level of brain edema was found in 60% of the members of the control group who had experienced the worst effect of jet lag. More studies are needed to clarify the connection of jet lag with minimal, subclinical brain edema.

In conclusion, these jet lag studies with 120 volunteers added another reason to take Pycnogenol® before and during flight, in addition to prophylaxis of thrombosis and edema of the lower limbs: Less jet lag.

Pycnogenol® and Hemorrhoids

Millions suffer, but it is estimated that only 20% visit a doctor. It's difficult to talk about the pain in the posterior. According to the National

Digestive Diseases Information clearinghouse, about 50% of the US population will have hemorrhoids by the age of 50. Even when in most cases hemorrhoids, come and go in a relatively short period of time, quality of daily life is considerably restricted during these weeks.

As we have mentioned, we often learn of new health benefits from Pycnogenol® users. When Professor Rohdewald first analyzed the French medical literature concerning Pycnogenol® for his physician friend, there were several anecdotal or small case reports that the pain, bleeding and itching due to hemorrhoids disappeared while taking Pycnogenol® supplements. These case reports were detailed, but uncontrolled. A clinical study was started in Italy to objectivate the role of Pycnogenol® in the treatment of hemorrhoids. (Ref. 11)

Painful to sit out: hemorrhoids

In this study, 84 subjects with an acute episode of external hemorrhoids were recruited in a doctor's office in Pescara, Italy. They were randomly assigned to receive either a Pycnogenol® supplement or identical placebo pills without active ingredients. The combination of internal and external treatment was also tested within this study. Other groups of patients received either a Pycnogenol® cream plus Pycnogenol® in tablets or the cream without Pycnogenol® plus placebo tablets. For treating these acute events, a high dose of Pycnogenol® (300 mg) was chosen for the first four days, followed by 150 mg for the following three days. (Ref. 11)

The placebo groups got the same number of tablets. The groups undergoing additional external treatment with the creams received the same number of Pycnogenol® as the first two groups. All groups were instructed how to change dietary habits to ameliorate the symptoms and to avoid aggravation of the hemorrhoids. Symptoms and signs of the hemorrhoids were assessed in weekly intervals.

The most visible sign, the hemorrhoid bleeding, disappeared completely within the period of investigation in the 3 Pycnogenol® groups, but was still present in the control group.

Table 10.5 Development of hemorrhoidal attacks (scores)

		Inclusion	7 days	14 days
Intravascular thrombus	1	2.3	1.9	1.7
	2	2.1	0.6	0.2
	3	2.3	0.3	0.2
	4	2.1	0.5	0.2
Severe pain	1	3.4	2.2	1.9
	2	3.2	0.8	0.2
	3	3.3	0.3	0.2
	4	3.1	0.7	0.2
Swelling	1	3.6	2.8	2.4
	2	3.6	1.1	1.1
	3	3.6	0.5	0.4
	4	3.5	1.2	1.0
Over-sensitivity	1	3.7	2.8	2.1
	2	3.8	1.4	1.4
	3	3.7	1.0	1.0
	4	3.7	1.3	1.3
Bleeding	1	2.8	1.3	1.0
	2	2.6	0	0
	3	2.7	0	0
	4	2.6	0	0

1) Placebo; 2) Pycnogenol® tablets; 3) Pycnogenol® tablets + Pycnogenol® cream; 4) Pycnogenol® tablets + basis cream

The scores for signs and symptoms decreased significantly in all Pycnogenol® groups compared to placebo. The combined treatment with Pycnogenol® tablets plus Pycnogenol® cream entailed a faster relief of signs and symptoms. The combination of Pycnogenol® with the basis cream was just a little bit more effective than Pycnogenol® alone.

More quality of life, less treatment costs

In parallel to the diminution of the hemorrhoids, the quality of life was improved for the Pycnogenol® groups, as indicated by work performance, walking, standing, and social activity. These improvements were reflected by treatment costs: 33–58% less for the Pycnogenol® groups compared to the control group, and by fewer work days lost: 4.6 versus 3 days.

These results advocate a treatment of hemorrhoids by a combination of internal and external application of Pycnogenol®. In this manner, the internal route may improve endothelial health and reduces thrombosis, the external application takes advantage of the wound healing and bacteriostatic activity of Pycnogenol®.

In another study by the Italian doctors, it was tested whether hemorrhoids, appearing during pregnancy, are disappearing within a shorter time when the young mothers were supplemented with 150 mg/day Pycnogenol® compared to an untreated control group. (Ref. 12) Following the 6 month period of supplementation, 75% of the women in the Pycnogenol® group were symptom free, whereas in the control group only 56% were free from hemorrhoids. Similarly, the edema of the lower legs, developed during pregnancy, disappeared after supplementation with Pycnogenol® 100 mg/day in 96.8% of the women after six months. (Ref. 12) The treatment with stockings was less successful as edema was still present at six months in 13.3% of this group.

Table 10.6 Restrictions in daily life due to hemorrhoids

		7 days	14 days
Social routine alterations	1	3.4	2
	2	2	1
	3	2	0.4
	4	2.2	1
Impairment in walking	1	3.3	2
	2	2	1
	3	1.2	0.6
	4	2	1
Impairment in standing	1	3.4	1.3
	2	2.2	0.4
	3	2	0.2
	4	2.3	0.5
Impairment in working performance	1	3.4	2.2
	2	2.5	1.1
	3	0.3	0.2
	4	2.6	1
Embarrassment, social withdrawal	1	3.7	2
	2	3.1	1.1
	3	2	0.8
	4	3	1
1) Placebo; 2) Pycnogenol® tablets; 3) Pycnogenol® tablets + Pycnogenol® cream; 4) Pycnogenol® tablets + basis cream			

Pycnogenol® and Tinnitus

Sometimes it is a ringing or a buzzing, or maybe a whistling or perhaps even a humming noise. It lasts for minutes, hours, even days – or longer. Tinnitus may come from a broad range of causes and mechanisms. It is believed that frequent exposure to loud noises is a cause. A frequent cause goes back to military service. According to the American Tinnitus Association, the Veteran's Administration will spend about 2.26 billion dollars in 2014 on disability payments for veterans suffering from tinnitus. (Ref. 13)

Unfortunately, there is no known standard therapy for tinnitus. Some "therapies" try to educate the patients how to cope with the disease. Other options are nerve stimulation or sound therapy, in severe cases a prostaglandin could be tried.

In Italy, doctors occasionally observed that when patients were treated with Pycnogenol® for their circulation problems, that their tinnitus also improved. These doctors had experience in treating tinnitus with drugs that increase microcirculation in the inner ear and they were willing to further test Pycnogenol® as a nutritional adjunct for tinnitus.

Tinnitus is reduced dose-dependently by Pycnogenol®

To test the effect of Pycnogenol® on a broader basis, patients with only one ear affected by tinnitus were recruited. (Ref. 14) In this study, 24 received 150 mg of Pycnogenol® per day, 34 got 100 mg daily and 24 remained untreated to serve as controls. Patients rated their symptoms using two different types of scores at the start of the study and at four weeks, which was the end of the study. The results are summarized in Table 10.7.

Table 10.7 Blood flow inside the ear

		Affected ear	
		Systolic	Diastolic
Pycnogenol® (100 mg)	Before	14.3	4.22
	After 4 weeks	21.2	8.23
Pycnogenol® (150 mg)	Before	13.2	3.2
	After 4 weeks	24.3	12.5
Controls	Before	13.6	4.12
	After	13.3	3.22

Tinnitus symptoms decreased to a greater extent with the higher dose, the control group remained unchanged.

By sophisticated ultrasound measurements, the investigators could demonstrate why Pycnogenol® produced a dose-dependent relief of symptoms: Pycnogenol® improved blood flow inside the inner ear. This is caused by the general effect of Pycnogenol® in improving blood flow by dilating blood vessels.

The higher dose produced a better blood flow than the lower dose.

This investigation indicates a dose-related effect of Pycnogenol® on the circulation inside the ear and the reduction of tinnitus. Pycnogenol® is an effective nutritional adjunct to improve inner ear circulation health.

Meniere's Disease

Meniere's disease is another dysfunction of the inner ear, associated also with tinnitus, but accompanied with dizziness or hearing loss, caused by a malfunction of the lymphatic flow inside the ear. In a study conducted at the Italian Chieti-Pescara University, researchers treated and monitored 107 patients between the ages of 35 and 55 who were diagnosed with Meniere's disease and suffering from symptoms like tinnitus. All patients were managed with best available management (BM), which included anticholinergics, benzodiazepines, antihistamines, corticosteroids, low salt diet and avoidance of caffeine, alcohol or other stimulants. In addition to BM treatments, the Pycnogenol® group was supplemented with 150 mg/day of Pycnogenol®, which was extremely helpful. (Ref. 15)

Results were recorded based on observational and reported scales for symptoms such as tinnitus, feeling of pressure and unsteady gait.

Inner-ear blood flow velocity was measured using a high-resolution, linear imaging probe. At baseline, flow velocity at the level of the affected ear was significantly lower in comparison with the other ear showing cochlear hypoperfusion.

After a three months period of treatment with Pycnogenol®, 45% of the patients were asymptomatic, in the untreated control group only 23%. Following the 6 month's intake of Pycnogenol®, percentage of asymptomatic patients raised to 87%, from the controls only 35% achieved a symptom-free status.

Over the course of six months, researchers also found Pycnogenol® to:

- Significantly improve inner-ear blood flow and reduce pressure as compared to control group (higher flow, higher diastolic component ($p < 0.05$)).

- Significantly improve patient-reported tinnitus as compared to control group ($p<0.05$ at three and six months).

- Reduce the number of missed work days due to inner-ear ailments as compared to control group ($p < 0.05$).

"The important effect of Pycnogenol® on improving microcirculation makes it a safe and natural option for those seeking relief from the symptoms of Meniere's disease, including tinnitus," said Dr. Gianni Belcaro, lead researcher of the study. "Because Pycnogenol® also has proven anti-inflammatory activity and antioxidant action, it may also help protect against the onset of tinnitus."

In this chapter, we have explored the nutritional adjunct to maintaining venous health and microcirculation, especially as it related to edema, jet lag, "economy class syndrome," hemorrhoids, and tinnitus. Now let's consider the many important special health benefits for women.

Chapter References

1. Gulati OP. Pycnogenol® in Chronic Venous Insufficiency and Related Venous Disorders. Phytother Res. 2013 Jun 15. doi: 10.1002/ptr.5019.

2. Arcangeli P. Pycnogenol® in chronic venous insufficiency. Fitoterapia 71(3): 236–244, 2000.

3. Koch R. Comparative study of Venostasin and Pycnogeno® in chronic venous insufficiency. Phytother Res 16: 1–5, 2002

4. Cesarone Mr, Belcaro G, Rohdewald P, et al. Comparison of Pycnogenol® and Daflon® in Treating Chronic Venous Insufficiency: A Prospective, Controlled Study. Clin Appl Thromb Hemost 12: 205–212, 2006

5. Riccioni C, Sarcinella R, Izzo A, et al. Efficacia della troxerutina associata al Pycnogenol® nel trattamento farmacologico dell'insufficienza venosa. Minerva Cardioangiol 52: 43–48, 2004

6. Belcaro G, Dugall M, Luzzi R, et al. Improvement of Venous Tone with Pycnogenol® in Chronic Venous Insufficiency: An Ex Vivo Study on Venous Segments. Int J Angiol 23: 47–52, 2014

7. Errichi BM, Belcaro G, Hosoi M, et al. Prevention of post thrombotic syndrome with Pycnogenol® in a twelve month study. Panminerva Med 53: 21–27, 2011

8. Cesarone MR, Belcaro G, Rohdewald P, et al. Improvement of signs and symptoms of chronic venous insufficiency and microangiopathy with Pycnogenol®. A prospective, controlled study. Phytomedicine 17: 835–839, 2010

9. Belcaro G, Cesarone MR, Rohdewald P, et al. Prevention of venous thrombosis and thrombophlebitis in long-haul flights with Pycnogenol®. Clin Appl Thromb Hemost 10: 373–377, 2004

10. Belcaro G, Cesarone MR, Steigerwalt RJ, et al. Jet-lag: Prevention with Pycnogenol®. Preliminary report: evaluation in healthy individuals and in hypertensive patients. Minerva Cardioangiol 56: 3–9, 2008

11. Belcaro G, Cesarone MR, Errichi B, et al. Pycnogenol® Treatment of Acute Hemorrhoidal Episodes. Phytother Res 24: 438–444, 2010

12. Belcaro G, Gizzi G, Pellegrini L, et al. Pycnogenol® in portpartum symptomatic hemorrhoids. Minerva Ginecol 66: 77–84, 2014

13. Caroll J. Defense Media Network

14. Grossi MG, Belcaro G, Cesarone MR, et al. Improvement in cochlear flow with Pycnogenol® in patients with tinnitus: a pilot evaluation. Panminerva Med 52: 63–67, 2010

15. Luzzi R, Belcaro G, Hu S, et al. Improvement in symptoms and cochlear flow with Pycnogenol® in patients with Meniere's disease and tinnitus. Minerva Med, 105: 245–254, 2014

Chapter Eleven | Women's Health

Pycnogenol® has special health benefits for women at all stages of life. In addition to the important health benefits to both sexes, Pycnogenol® helps women through difficult periods, pregnancy, and most important, menopause.

Less Period Pain with Pycnogenol®

It is the monthly burden for most women of reproductive age – some experience mild discomfort before menstruation (PMS), many have to take analgesics to be able to work. For some, the pain can seriously affect the quality of life. The complete replacement of tissue lining the uterine cavity, the endometrium, during the menstrual period represents a wound healing process and involves inflammatory processes. The inflammation is initiated by prostaglandins, developing during menstruation, causing uterine contractions and pain.

In medical terminology, this is known as dysmenorrhea

The prevalence of dysmenorrhea is highest in adolescent women, with estimates ranging from 20% to 90% depending on the diagnostic standards applied. For women dysmenorrhea is the most common reason for absence from work.

Being informed about the anti-inflammatory activity and the positive influence on wound healing of Pycnogenol®, two Japanese gynecologists started a pilot trial. They gave 30 mg of Pycnogenol® daily to 39 women seven days before the "bad days." Both abdominal pain and cramps were clearly alleviated, according to the judgment of the women. (Ref. 1)

The doctors, now Pycnogenol® believers, investigated in a new study the effect of 60 mg pine bark extract over a period of two cycles. Compared to pretreatment, the scores for abdominal pain shifted from high pain to low pain, the women reported less days with menstrual pain. Consequently, their use of analgesics dropped. (Ref. 2)

A multi-center study demonstrates less need of pain killers in dysmenorrhea

A multi-center study with 116 women in 4 Japanese hospitals showed that regular supplementation with 60 mg daily of Pycnogenol® halved their use of analgesics. Their need for analgesics remained at that low level even after stopping their Pycnogenol® supplements. (Ref. 3)

The placebo group also reduced their consumption of analgesics, but not to the same extent. Women had 50% fewer painful days in the Pycnogenol® group, while no change in this respect was reported in the placebo group.

Less Pain in Endometriosis

Endometriosis is another painful event associated with the female menstruation cycle. Cells lining the uterine cavity, the endometrium, are normally shed during menstruation and the tissue gets renewed. In endo-

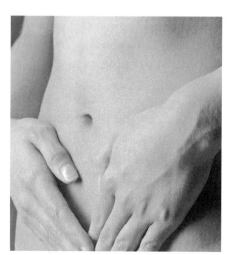

metriosis, cells displaced from the endometrium, migrate outside and to other organs and grow there as endometrioma and still respond there to the cycle. Because these displaced cells cannot leave the host organs, painful inflammation erupts.

The standard treatment of endometriosis involves anti-inflammatory drugs. More advanced cases require surgery. Hormonal treatment, to block estrogen production, with the synthetic peptide Leuprorelin is another

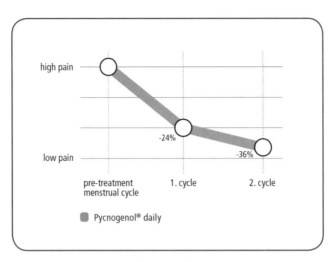

Figure 11.1. Pycnogenol® gradually decreases menstrual pain. (Ref. 2)

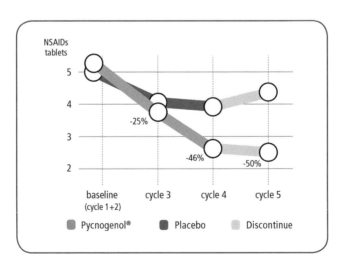

Figure 11.2. Women supplemented with Pycnogenol® required less pain medication. (Ref. 3)

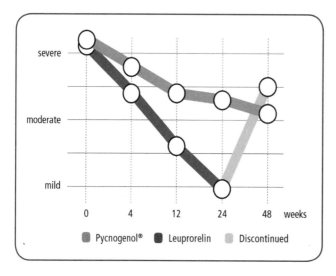

Figure 11.3. Pycnogenol® lowers pelvic pain in endometriosis. (Ref. 4)

option. Leuprorelin has to be injected as a long-lasting deposit under the skin. It prevents menses and also conception. However, treatment with Leuprorelin has to be limited to six months to avoid osteoporosis and other unwanted effects that occur if treatment is stopped too late.

In Japan, a study of 58 women suffering from endometriosis compared the relief of menstrual pain by taking either the drug Leuprorelin daily or 60 mg of the nutritional supplement Pycnogenol®. Leuprorelin, which suppresses menses, reduced the pain during the 6 months treatment period from severe pain to mild pain. However, 24 weeks after the obligatory end of the treatment, pain came back again, ranging from moderate to severe. (Ref. 4)

Pycnogenol® alleviated pain more slowly from severe to just moderate, however, the reduction of pain persisted over the following 2 weeks. It did not interfere with the menstrual cycle and estrogen level. Six of the women taking Pycnogenol® left the study because they became pregnant.

A specific antigen indicated that the size of the endometrioma was shrinking due to supplementation with Pycnogenol®. This investigation advocates the supplementation with Pycnogenol® as an alternative to hormonal treatment of endometriosis. (Ref. 4)

A trial from Brazil tested the effect of a combination of oral contraceptives in combination with 100mg Pycnogenol® on endometriosis. Oral contraceptives diminish the pain caused by endometriosis, but the addition of Pycnogenol® helped 54–57% of the women to be completely pain free. This could not be achieved with the oral contraceptives alone. (Ref. 5)

Biopsies, taken from the women with endometriosis, revealed a lower concentration of the vascular endothelial growth factor and of aromatase inside the endometrial tissue when patients had been treated with the combination of Pycnogenol® with the oral contraceptives. (Ref. 6) Both agents are involved in the painful process of endometriosis.

Less Pain During Pregnancy

During the last 3 months of pregnancy, many women complain of pain in the lower back, hip joint, pelvis or calf cramps. In a study with 80 pregnant women, supplementation with 30 mg of Pycnogenol® daily until delivery resulted in a highly significant decrease of all types of pain, starting after two weeks. No significant alleviation of pain was reported in the control group. (Ref. 7)

No unwanted effects were observed in the Pycnogenol® group during treatment. Normal deliveries occurred with 68 women in the group and 12 women delivered via Cesarean section.

The gynecologists conclude, "Supplementation with Pycnogenol® in the third trimester of pregnancy seems to offer a safe and effective way of alleviating pregnancy-associated pain."

Even though animal studies show there are no teratogenic effects (causing birth defects) due to Pycnogenol®, we advise women in their first trimester of pregnancy to not take Pycnogenol® – as a general precaution.

Relief from Menopausal Symptoms

It is a normal burden for women between 45 and 60 to have menopausal problems. There are the hot flashes, night sweats and gynecological troubles, sometimes mood swings from depression, anxiety and fatigue to nervousness and over-activity. Many women have difficulty in sleeping, and others complain of painful sensations. From our previous experiences we could predict that Pycnogenol® would bring relief from some of the symptoms. Pycnogenol® attenuates pain in dysmenorrhea and pregnancy, it has a positive influence on cognitive functions and it normalizes high blood pressure. However, Professor Rohdewald wanted to have a good study conducted to ascertain if Pycnogenol® would perform as expected.

Every menopausal symptom improved

With the expectation that Pycnogenol® would bring some benefit for women in the period of their menopause, a study was started at a department of gynecology in Taiwan. A comparison was made between 200 mg Pycnogenol® daily and a placebo in a double-blind manner over a period of 6 months. At the beginning of the study, 155 women between 45 and 55 years of age answered questions from a standard gynecological "Women's Health Questionnaire" to describe their frequency of menopausal symptoms and levels of discomfort. (Ref. 8)

Every symptom listed in the questionnaire steadily decreased in the Pycnogenol® group during the six months from "clear discomfort" or "little discomfort" towards "little discomfort" or "no discomfort." Pycnogenol® was significantly superior to placebo, which had little or no effect.

Remarkably, the menopausal symptoms of the Taiwanese women showed a pattern different from European women. Whereas the Europeans suffer predominantly from hot flashes and night sweats, the Taiwanese were complaining most frequently of somatic troubles such as tiredness, sleep problems, headache, and anxiety.

154

Table 11.1 Change in menopausal symptoms (Ref. 8)
Increasing number signalize better conditions

	Pycnogenol®		Placebo	
	Enrollment	6 months	Enrollment	6 months
Feelings				
Miserableness and sadness	3.15	3.40	3.11	3.07
Irritability	2.46	3.12	2.50	2.64
Worry about growing old	2.50	3.17	2.36	2.52
Reduced well-being	2.85	3.23	3.03	2.95
Sleep				
Tiredness daytime	2.07	3.04	1.98	2.28
Restlessness	2.65	3.16	2.76	2.78
Early morning wakening	2.50	3.14	2.52	2.48
Difficulty falling asleep	2.59	3.61	2.41	2.41
Anxiety				
Panicky feelings	3.03	3.30	3.08	3.10
Palpitation	2.55	3.17	2.76	2.74
Feel tense/wound up	2.60	3.14	2.78	2.58
Menstrual				
Breast tenderness	3.09	3.19	3.05	2.91
Heavy bleeding	2.59	3.16	2.50	2.50
Bloatedness	2.71	3.30	2.75	2.93
Vasomotor				
Hot flashes	3.37	3.86	3.20	3.25
Night sweats	3.38	3.65	3.48	3.43
Pain				
Headaches	2.20	3.13	2.63	2.73
Backache/pains in limbs	2.26	3.16	2.33	2.28
Pain and needles in hands & feet	3.10	3.40	3.00	3.12
Sexual				
Loss of sexual interest	2.67	3.21	2.51	2.65
Vaginal dryness	2.41	3.18	2.61	2.42
Attractiveness				
Not lively	2.18	3.05	2.39	2.50
Feeling unattractive	2.35	3.13	2.50	2.61
Memory				
Difficulty in concentrating	2.36	3.10	2.61	2.36
Poor memory	1.93	3.00	2.02	2.33

At first, it seems to be a bit astonishing that every symptom, especially the somatic symptoms, was gradually improving, not dramatically, but consistently. However, it was known from previous studies that Pycnogenol® improves memory and learning. And it was able to attenuate hyperactivity and to stimulate attention.

From these observations one can accept that loss of memory, one of the menopausal symptoms, should be positively influenced. The reason for the improvement in brain function could be a stimulation of neurotransmitters by nitric oxide (NO) as we have discussed in earlier chapters. Besides memory, sleep behavior is also influenced by NO, which is produced in larger quantities with the aid of Pycnogenol®. As the volunteers reported fewer sleep problems, they were most probably in better condition in the morning, starting the day with less anxiety and better self-confidence. In turn, they may feel themselves more attractive and are less depressed.

As a positive "by-product" of the study, it was found that the Taiwanese women had a slightly lower blood pressure and lower blood cholesterol levels. In addition, the antioxidant status in blood was significantly improved.

An additional study proves amelioration of menopausal symptoms

For confirmation, a very similar second study was performed in Italy, but using half the amount of Pycnogenol® given to the women in the Taiwanese study. In this Italian study, a total of 65 women between 40 and 50 years of age were divided into two groups. One group of 33 women took 100 mg of Pycnogenol® daily, the other 32 women became the control group. (Ref. 9) At the start of the study and at eight weeks later, all participants answered the questions from a standard Menopause Symptoms Questionnaire. At the start of the study, all women were instructed on how to control or limit menopausal symptoms through a healthier lifestyle. Following the eight weeks, all symptoms in the Pycnogenol® group improved compared to what they were experiencing at the start of the study. This trend was not seen in the control group. In contrast to the results obtained with the Taiwanese women, sleep problems were rarely reported in this study at start. The most prominent symptoms were irregular periods, hot flashes, night sweats, fatigue, bloating, irregular heartbeat, irritability, headaches and digestive problems. Relative to the control group, the amelioration of hot flashes, bloating, irregular heartbeat and digestive problems became significant, whereas the improvement of the other symptoms failed to reach the level of significance.

So the European population presented another pattern of symptoms, and, as one could expect, the cardiovascular troubles responded well to supplementation with Pycnogenol®. A surprising finding was the good results for bloating and digestive problems.

The antioxidant activity in blood of the Pycnogenol® group was also improved as in the Taiwanese study.

Third menopause study

A third study was done in Japan with an even smaller dose of 60 mg Pycnogenol® daily, given for 12 weeks to 79 women. The control group consisted of 77 women who received a placebo. (Ref. 10) The women, age between 42 and 58, answered questions from the Women's Health Questionnaire at the study start and again after four and twelve weeks. All menopausal signs and symptoms were improved relative to start in the Pycnogenol® group. The vasomotoric symptoms, as hot flashes, were significantly improved versus a placebo after 4 and 12 weeks. The women in the Pycnogenol® group also reported significantly fewer sleeping problems. The symptom "feeling tired and worthless" reached statistical significance only after 12 weeks. A couple of symptoms remained at a borderline statistical significance, due to the high placebo effect.

The evaluation of the results using the Kupperman Index showed the overall advantage of supplementation with Pycnogenol® versus placebo. This Index summarizes 15 menopausal symptoms.

Clinical conclusions

The results of the three studies suggest that Pycnogenol® is an alternative to hormone replacement therapy that has some unwanted effects. The results from the study with 200 women are tabulated in Table 11.1 (Ref. 8) Pycnogenol® makes the menopause more tolerable and even pleasant for many: Less symptoms, better sleep, feeling more attractive, hence in good mood. And the volunteers tolerated the supplementation with Pycnogenol® very well.

Further good news is that experiments with mice suggest that Pycnogenol® can attenuate osteoporosis during menopause. Mice, used as a model for osteoporosis, received Pycnogenol® in their drinking water. Compared to an untreated control group, the mice getting Pycnogenol® had stronger bones, a lower bone turn-over, a lower activity of bone resorption and a higher bone density. (Ref. 11) These findings suggest that Pycnogenol® could be able to counteract osteoporosis. Please review Chapter Four for more details on Pycnogenol® and bone health. Pycnogenol® can be considered a reasonable option for getting through menopause in better shape and with a good mood. Also, considering that since Pycnogenol® improves endothelial function in the arteries, the increased risk of cardiovascular disease associated with post-menopause can be lessened as well. Pycnogenol® becomes an option with more than one purpose for menopausal women.

Now let's look at some of the health benefits specific for men in the next chapter.

Chapter References

1. Kohama T, Suzuki N. The treatment of gynecological disorders with Pycnogenol®. Eur Bull Drug Res 7: 30–32, 1999

2. Kohama T, Suzuki N, Ohno S, Inoue M. Analgesic efficacy of French maritime pine bark extract in dysmenorrhea. An open clinical trial. J Reprod Med 49: 828–832, 2004

3. Suzuki N, Uebaba K, Kohama T, et al. French Maritime Pine Bark Extract significantly lowers the requirement for analgesic medication in dysmenorrhea. A multi-center, randomized, double-blind, placebo-controlled study. J Reprod Med 53: 338–346, 2008

4. Kohama T, Herai K, Inoue M. Effect of French Maritime Pine Bark Extract on endometriosis as compared with Leuprorelin acetate. J Reprod Med 52: 703–708, 2007

5. Maia H, Haddad C, Casoy J. Combining oral contraceptives with a natural nuclear factor-kappa B inhibitor for the treatment of endometriosis-related pain. Int J Women's Health 6: 35–39, 2014

6. Maia H, Haddad C, Pinheiro N, Casoy J. The Effect of oral contraceptives combined with Pycnogenol® (Pinus Pinaster) on aromatase and VEGF expression in the eutopic endometrium of endometriosis patients. Gynecol Obstet (Sunnyvale) 4:203, 2014 – doi: 10.4172/2161–0932.1000203

7. Kohama T. Nutritional supplements in clinical practice. Progr Med 24: 1503–1510, 2004

8. Yang HM, Liao MF, Zhu SY, et al. A randomized, double-blind, placebo-controlled trial on the effect of Pycnogenol® on the climacteric syndrome in perimenopausal women. Acta Obstet Gynecol Scand 86: 978–985, 2007

9. Errichi S, Bottari A, Belcaro G, et al. Supplementation with Pycnogenol® improves signs and symptoms of menopausal transition. Panminerva Med 53: 65–70, 2011

10. Kohama T, Negami M. Effect of low-dose French maritime pine bark extract on climacteric syndrome in 170 perimenopausal women. J Reprod Med 58: 39–46, 2013

11. Mei L, Mochizuki M, Hasegawa N. Protective effect of Pycnogenol® on ovariectomy-induced bone-loss in rats. Phytother Res 26: 153–155, 2012

Chapter Twelve | Pycnogenol®
for Men

It would not be fair to men for us not to say something about how Pycnogenol® has health benefits specifically for them, beyond the benefits to both sexes. However, these benefits are mostly to sperm health (which of course is also important to women), erectile function, and to athletic performance (which applies to women as well, but men often seem to be more concerned with their athletic performance).

Improvement of Sperm Quality with Pycnogenol®

It is estimated that 60% of couples, suffering from futile attempts for conception, are seemingly infertile because one or more sperm parameters are not normal. As there is some evidence that free radicals may damage sperms, it was logical to try to fight the basis of infertility with a powerful scavenger of free radicals. An American specialist for reproductive endocrinology tested the effect of Pycnogenol® on 18 men with abnormal sperms.

Following the 90 days of supplementation with 200 mg Pycnogenol® daily, the shape of the sperms became normalized in 38% of the men. (Ref. 1) The binding of the sperms to a simulated surface of a human egg cell was improved by 19%.

The increase in quantity of normal functioning sperms may allow men to overcome subfertility.

Erectile Health and Prelox®

Normal erectile function depends heavily on nitric oxide (NO) availability. Penile erection requires the relaxation of the cavernous smooth muscle, which is triggered by NO. For those men having NO production within the low-normal range, supplementation with Pycnogenol® may help keep the NO production in the more optimal normal range. This is a nutritional factor and not as powerful as Erectile Dysfunction (ED) pharmaceuticals, but still, it may help many men.

In 2003, Bulgarian physicians at the Seminological Laboratory SBALAG in Sofia, Bulgaria gave Pycnogenol® supplements because of its ability to increase production of NO by nitric oxide synthase, along with the amino acid dietary supplement L-arginine, which is the substrate for this enzyme. The study included 40 men, aged 25–45 years, without confirmed organic erectile dysfunction. Throughout the 3-month trial period, volunteers took 1.7 g L-arginine per day in the form of L-arginyl aspartate. During the second month, patients were additionally supplemented with 40 mg Pycnogenol® two times per day; during the third month, the daily dosage was increased to three 40-mg Pycnogenol® tablets. (Ref. 2)

After one month of treatment with L-arginine two patients (5%) experienced a normal erection. Treatment with a combination of L-arginine and Pycnogenol® for the following month increased the number of men with restored sexual ability to 80%. Finally, after the third month of treatment, 92.5% of the men experienced a normal erection. The researchers concluded, "Oral administration of L-arginine in combination with Pycnogenol® causes a significant improvement in sexual function in men with ED without any side effects."

In 2010 at the Gabriele D'Annunzio University in Chieti-Pescara, Italy, a double-blind, placebo-controlled study assessed the effects of Pycnogenol® in 124 patients (aged 30–50 years) with moderate ED over an investigational period of 6 months. (Ref. 3)

This study assessed the effects of supplementation of a combination of Pycnogenol® and L-arginyl aspartate in 124 patients (aged 30–50 years) with moderate ED over an investigational period of 6 months. The International Index of Erectile Function (IIEF) was used to quantify changes in sexual function.

The erectile domain of the IIEF improved with the Pycnogenol®/arginine combination supplement from a baseline mean score of 15.2 to 25.2 after three months and 27.1 after six months. In the placebo group there was an increase from a baseline score of 15.1 to 19.1 and 19.0 after three and six months, respectively. The effects of the supplement were statistically significant compared with placebo (P < 0.05). Mean total plasma testosterone levels increased significantly from 15.9 to 18.9 nmol/L (P < 0.05) after six months with the supplement, compared to an increase from 16.9 to 17.3 nmol/L in the placebo group.

In 2012 in Osaka, Japan, a clinical study was conducted in volunteers with mild to moderate erectile dysfunction to investigate the efficacy of Pycnogenol® and the amino acid L-arginine. (Ref. 4)

In this double-blind, placebo-controlled study, subjects were instructed to take a supplement containing Pycnogenol® 60 mg/day, L-arginine 690 mg/day and aspartic acid 552 mg/day, or an identical placebo for 8 weeks. The results were assessed using the five-item erectile domain (IIEF-5) of the International Index of Erectile Function. Additionally, blood biochemistry, urinalysis and salivary testosterone were measured. Eight weeks of supplement intake improved the total score of the IIEF-5. In particular, a marked improvement was observed in 'hardness of erection' and 'satisfaction with sexual intercourse.' A decrease in blood pressure, aspartate transaminase and γ-glutamyl transpeptidase (γ-GTP), and a slight increase in salivary testosterone were observed in the supplement group. No adverse reactions were observed during the study period. The researchers concluded, "Pycnogenol® in combination with L-arginine as a dietary supplement is effective and safe in Japanese patients with mild to moderate erectile dysfunction."

What is Prelox®?

Prelox® is a patented combination of two potent ingredients clinically proven to increase a man's ability to achieve and sustain an erection: Pycnogenol® and L-Arginine. Nitric Oxide (NO) plays an important role with dilation of blood vessels. Healthy blood flow to the genital area is key for sexual satisfaction. Prelox® is a proprietary blend which has been shown in various studies to increase NO production. This combination has a beautifully synergistic effect on blood flow into the vessel. Prelox® stimulates circulation and blood flow, it protects blood vessels from the damage that can occur from normal aging, and enhances responsiveness, sexual stamina and enjoyment. Prelox® is marketed under Horphag's licence by selected marketers worldwide and Prelox® is protected by US patent 6.565,851 and other international patents.

Chapter References

1. Roseff SJ. Improvement in sperm quality and function with French maritime pine tree extract. J Reprod Med 47: 821–824, 2002

2. Stanislavov R, Nikolova, V. Treatment of erectile dysfunction with Pycnogenol® and L-arginine. J Sex Marital Ther 29:207–13, 2003

3. Ledda A, Belcaro G, Cesarone MR, et al. Investigation of a complex plant extract for mild to moderate erectile dysfunction in a randomized, double-blind, placebo-controlled, parallel-arm study. BJU Int 106:1030–1033, 2010

4. Aoki H, Nagao J, Ueda T, et al. Clinical Assessment of a Supplement of Pycnogenol® and L-arginine in Japanese Patients with Mild to Moderate Erectile Dysfunction. Phytother Res 26: 204–207, 2012

Chapter Thirteen | Children's Mental Health

Just as in the situation with adults, Pycnogenol® is a nutritional supplement that helps supply children with beneficial nutrients that may be in short supply in the typical child's diet. The normal health benefits for children include helping to keep inflammation and oxidative stress within normal bounds. A daily supplementation with 1 mg/kg/ day is used for children between five and fourteen years in a healthy practice. Some children may find that greater Pycnogenol® intakes are beneficial. Examples include children with allergies and hyperactivity. In this chapter we will discuss Attention Deficit Disorder (ADD) in children, but the information also applies to adults.

Pycnogenol® and Attention Deficit Disorder (Children and Adults)

Attention Deficit Disorder (ADD) and Attention Deficit Hyperactivity Disorder (ADHD) are a group of behavioral problems that used to be called "hyperactivity." They involve impulsive behavior, the inability to keep focused on a task, and/or hyperactivity. We learned quite unexpectedly that Pycnogenol® is a nutritional adjunct that helps in these disorders as well.

Attention Deficit Disorder (ADD) involves the inability to keep focused on a task, impulsive behavior and/or hyperactivity. ADD is much broader than the excessive physical activity of hyperactivity; it includes behavioral and mental disorders that keep one from learning or performing well, even though the individual had the mental capability to do so.

ADD affects about five-to-ten percent of school-aged children in the US, and is the basis for about one-half of the childhood referrals to diagnostic clinics. ADD is seen ten times more frequently in boys than girls.

The cause of ADD is not known, but structural abnormalities have been ruled out by CAT, MRI and EEG scans. The leading suspect appears to be neurotransmitter abnormalities, possibly associated with decreased activity or stimulation in the upper brain stem and frontal midbrain. There is also suspicion that toxins, environmental problems or neurologic immaturity could be involved.

The American Psychiatric Association lists 14 signs of which at least eight must be present to be officially classified as ADD. These fourteen signs are:

(1) often fidgets with hands or feet, or squirms in seat (restlessness),
(2) has difficulty remaining seated when required to do so,
(3) is easily distracted by extraneous stimuli,
(4) has difficulty awaiting turn in games or group activities,
(5) often blurts out answers before questions are completed,
(6) has difficulty in following instructions,
(7) has difficulty sustaining attention in tasks or play activities,
(8) often shifts from one uncompleted task to another,
(9) has difficulty playing quietly,
(10) often talks excessively,
(11) often interrupts or intrudes on others,
(12) often does not seem to be listening to what is being said,
(13) often loses things necessary for tasks or activities,
(14) often engages in physically dangerous activities without considering possible consequences.

Different combinations of inattention or impulsiveness or hyperactivity constitute different sub-groups of ADD. Note that ADD is **not** retardation, although learning under these conditions is difficult.

Not Just Kids

Usually, ADD begins by four-to-seven years of age, with the peak ages in which treatment is sought is between eight-to-ten years of age. However, adolescents and adults also suffer from ADD, with the estimated incidence ranging from thirty-to-seventy percent of the affected children also being ADD adults. In adults, ADD is rarely a case of "adult onset," but an abating childhood condition. The physical "hyperactivity" lessens with age, but the adult still has marked attention problems and can be impulsive. Problems in the adolescent and adult occur predominantly as academic failure, low self-esteem, and difficulty learning appropriate social behavior. Often they have personality-trait disorders, anti-social behavior, short attention spans, and are impulsive and restless, and have poor social skills.

Conventional Treatment Problems

Conventional treatment is with central nervous system stimulants such as amphetamines. Ritalin (methylphenidate or methyl-alpha-phenyl-2-piperidineacetate) is widely prescribed with the result of five-to-ten percent of our youngsters going to school drugged. According to the Merck Manual, common side effects of Ritalin are sleep disturbances (e.g. insomnia), depression or sadness, headache, stomachache, suppressed appetite, elevated blood pressure, decreased learning, behavioral changes and reduction of growth.

An Interesting Letter

In February 1995, Dr. Passwater received a letter from a Valdosta, Georgia elementary school teacher for seventeen years who at the time was the mother of five children including an ADD daughter. We emphasize that she was a teacher because teachers know what normal behavior in the classroom is. The mother reported that she gave Pycnogenol® to her daughter and had been able to take her completely off Ritalin. She was concerned about safe dosage limits and said she would appreciate any information regarding ADD and Pycnogenol®. At that time, Dr. Passwater had nothing unique to tell her about ADD and Pycnogenol®.

In April 1995, Dr. Passwater received a call from his friend, Myles Lipton of St. Louis, who wanted to know if he had any information on ADD and

Pycnogenol®. Mr. Lipton knew of several friends who belonged to an ADD Support group, and they were markedly improved.

As we have mentioned several times, we often get reports of how Pycnogenol® helps this or that, but these "studies of one" are considered as anecdotal reports and have little scientific value. But, when you get enough smoke, you begin looking for the fire.

At that time, ADD support groups were somewhat of a rarity. ADD sufferers aren't fond of meetings, and it is difficult to conduct such meetings. Dr. Passwater suggested to Mr. Lipton that he obtain some objective data since he was familiar with several of the group members and the progress that was being made. Mr. Lipton designed a questionnaire and contacted members of the group. Excerpts follow here.

Study Excerpts

Interviews were conducted at the ADD Support Group meeting on April 1995. Dan H., a thirty year old male always had trouble concentrating and had not performed well in school as a child. He was always very agitated. When asked why he was taking Pycnogenol®, he responded. "I attended ADD meetings for three months, and in speaking with other members, I heard lots of complaints of side effects of various drugs that their doctors had prescribed. I did not want to take these drugs. I was talking with a guy at work and he gave me an audio tape of a lecture by Dr. Lamar Rosquist. I listened to the tape and there was a small segment in which Dr. Rosquist mentioned that a little girl with ADD was helped by taking Pycnogenol®.

Dan started taking Pycnogenol®. The first day he took 300 mg and then dropped to 150 mg daily. Within days he noticed that he was thinking clearer. His thinking got sharper and continued to improve for a week and then leveled off. He used to be hyper, but was now calmer and more alert to what was going on about him. He reported being less depressed and has less of a sense of being anxious.

Dan reported, "How I have felt in the last two months has been the best thing that has ever happened to me. I am undoing old bad habits, the worst of which was a lot of procrastination in taking care of things that I needed to do. I do a better job of listening to people and when they talk to me, my mind does not drift off anymore."

Jill

Jill G. was a 36 year-old female. She had a short attention span, a short-fused temper, was always fidgety, impatient, and a master procrastinator.

When asked why she started taking Pycnogenol®, she responded. "Dan, a member of our group came to a meeting and seemed much different. He used to never sit still, interrupt others frequently, and his eyes always seemed like they were popping out of their sockets. But this time, he was sitting still and attentively listening to others, and his eyes looked normal. I asked him what was up and he told me that he was taking 150 milli-grams of Pycnogenol® a day."

Jill began taking 150 milligrams of Pycnogenol® daily and noticed a significant improvement in about two weeks. At first there was an improvement in bowel function. [The drug Zoloft may make one constipated, and she had cut back on it.] She noticed a major improvement in the dark circles under her eyes. She seemed to have more motivation and did not procrastinate as much.

Jill adds, "I had been taking the antidepressant Zoloft and have gotten off of it with consultation with my doctor. I have noticed that my husband Jeff, who also has ADD, has had major improvements and a noticeable increase in confidence since he has been taking Pycnogenol®."

Jeff

Jeff G. (husband of Jill) was a 35 year-old male who had always been compulsive, had a lot of trouble concentrating on things and was always losing things.

When Jeff was asked why he was taking Pycnogenol®, he responded, "Because of the enormous positive behavioral changes in Dan H., a member of our group."

Jeff began taking 180 milligrams of Pycnogenol® daily at the same time his wife started. He noticed a marked improvement in about three hours! He went to the bar that he frequents but always feels uncomfortable in. But after three hours of taking Pycnogenol®, he felt more at ease. As time went by, the improvements increased, he had less anxiety feelings, more energy, was calmer, and had better relationships with fellow employees at work.

Jeff added, "I had been taking medication that had the side effect of increasing nervousness, and Pycnogenol® has allowed me to reduce the drugs and have an overall better feeling."

David

Thirty-year old David L. was hyperactive all through school and had problems concentrating. David began taking 75 milligrams of Pycnogenol® daily soon after he noticed the changes in Dan. David noticed positive results after one week. He had less anxiety, felt calmer and was thinking more clearly.

When asked if he had anything to add, David said, "Just that I really feel better since taking Pycnogenol® and I hope that the positive results continue."

Laura

Let's leave the ADD group and return to the Georgia teacher's daughter for an update. Laura D. was 12 years old. She had been diagnosed as having ADHD and Oppositional and Defiant Behavior Disorder. She had been taking 70 milligrams of Ritalin and 50 milligrams of Tofranil daily. Within 30 days of starting both Pycnogenol® and a combination antioxidant formula, her parents noted improvements in both behavior and attention span to a greater extent than that realized with the prescription drugs. Additional improvements in appetite and personality traits were noted.

Beth D., Laura's mother told Dr. Passwater, "As a parent, I would never revert to the prescription drug Ritalin. As a mother of five children and an elementary teacher of 17 years, I have had a wealth of experience in working with children with ADD/ADHD. Most of these children have been on various prescription drugs of which Ritalin, in various dosages, being the most common. It appears that the use of prescription drugs for this disorder has become common place to the point that it is over prescribed. The long-term side effects or ramifications of using drugs like these described are still unknown. The ability to use a food supplement and obtain like, positive results must be pursued."

Dr. Passwater mentioned the small study by Mr. Lipton as part of his presentation at the Second International Symposium on Pycnogenol® in May 1995 in Biarritz, France.

In August 1995, Dr. Passwater received a letter from the psychologist who led the ADD group, Dr. Julie Pauli. She suffered from ADD herself. She, herself, had good experiences with Pycnogenol®. Personally, she could concentrate better, needed less caffeine, took less antidepressant and her "inner world" was now organized and her emotional reactivity was reduced. She said that her psychologist associate, Dr. Stephen Tennenbaum, had similar results.

She asked if Pycnogenol® had been studied as an adjunct treatment for ADD and, if not, suggested that studies be done. Dr. Passwater proposed she write to Professor Rohdewald.

Professor Rohdewald suggested that the two psychologists make a presentation of their experiences with Pycnogenol® and ADHD. Professor Rohdewald observed that both speakers had difficulties in presenting their findings, because both of them had ADHD. Nevertheless, the presentation was well organized and instructive.

A Useless Study

Professor Rohdewald wanted to get more fundamental information about Pycnogenol and ADHD and suggested that under the lead of these psychologists, a double-blind, clinical study was initiated comparing Pycnogenol®, placebo, and Ritalin. The outcome of this study was that no difference between placebo and Ritalin, no difference between Pycnogenol® and Ritalin and no difference between placebo and Pycnogenol® was found. As the study could not differentiate between the doubtless active drug, Ritalin, and the doubtless inactive placebo, the study was a flop. As the study did not have the statistical power to detect differences between active drugs and inactive placebo, no conclusions could be drawn.

Positive Reports Again from USA

Thereafter, in 1999, news came again from the USA. An MD recommended Pycnogenol® for treatment of ADHD in a booklet, based on her positive experiences with children. (Ref. 1) Another MD reported the outcome of a classical test: A boy with ADHD got Pycnogenol® in addition to a prescribed drug. After 2 weeks, his parents noted improvement of symptoms. Then intake of Pycnogenol® was stopped to compare the effects of the stimulant plus Pycnogenol® with stimulant alone. Within 2 weeks of stopping Pycnogenol®, the boy became more hyperactive and impulsive, marked by demerits and physical altercations in school. Pycnogenol® was given again and within 3 weeks symptoms improved again. (Ref. 2)

A Japanese Study

A year later, a neurologist reported in a Japanese newspaper that 40 children with ADHD did better in school after supplementation with Pycnogenol® with a success rate of 70%. (Ref. 3) This investigation was not sponsored by anybody, so we wondered why the doctor did this study. The answer was astonishing to us. As Pycnogenol® had been very helpful for him and his wife against painful symptoms of "frozen" shoulder, he got generally interested in the power of Pycnogenol® and collected literature about Pycnogenol®. He came across the reports about ADD from the USA and decided to find out whether it worked in Japanese children, too. Unfortunately, he was neither willing to publish his findings in a scientific journal nor to do a new, controlled study.

All together, these reports gave Professor Rohdewald a reason to start again a controlled clinical investigation.

The Clinical Proof of Benefits for Children with ADHD

The study with 61 children was done at a university hospital in Bratislava, Dept. of Child Psychiatry. The double-blind study, published in 2006. (Ref. 4), demonstrated that 1 mg/kg/ day Pycnogenol® was better than a placebo in relieving hyperactivity and improving attention and coordination in children with ADHD. The judgments were made by teachers and parents.

Interestingly, some parents realized during the study that there was no progress at all with their children while others behaved better and better. So they blamed the principal investigator for a useless treatment of their children, as they felt that their kids were most probably being given the placebo. The double-blind study was stopped for this ethical reason several weeks before planned. The result still showed that Pycnogenol® significantly improved behavior, but now, at least, the children, formerly receiving the placebo, could get finally Pycnogenol® without waiting for the planned end of the investigation.

Besides the clear improvement of ADHD, the investigation of biochemical markers during this study showed that children with ADHD produce more stress hormones (Ref. 5), suffer more from oxidative stress and, that the oxidative stress finally damages the DNA of these kids. (Ref. 6) All those effects of ADHD were be significantly reduced by Pycnogenol®.

The data demonstrate that Pycnogenol® helps people manage the symptoms of ADHD. It is suggested that Pycnogenol® may act by reducing the production of stress hormones, or by interference with the system of neurotransmitters. The question about the exact mechanism of action against ADHD remains open.

Our picture now, composed from testimonials, anecdotal reports and investigations on a scientific basis, shows the Pycnogenol® offers a natural nutritional adjunct or alternative to the stimulant Ritalin.

Chapter References

1.   Hanley JL. Attention Deficit Disorder. Impact Communications Inc., Green Bay, WI, USA, 17–19, 1999

2.   Heimann SW. Pycnogenol® for ADHD? J Am Acad Child Adolesc Psychiatry 38: 357–358, 1999

3.   Masao H. Pycnogenol®'s therapeutic effect in improving ADHD symptoms in children. Mainichi Shimbun, 2000, Oct. 21

4.   Trebaticka J, Kopasova S, Hradecna Z, et al. Treatment of ADHD with French maritime pine bark extract, Pycnogenol®. Eur Child Adolesc Psychiatry 15: 329–335, 2006

5   Chovanova Z, Muchova J, Sivonova M et al. Effect of a polyphenolic extract, Pycnogenol®, on the level of 8-oxoguanin in children suffering from ADHD. Free Radic Res 40: 1003–1010, 2006

6   Dvorakova M, Jezova D, Blazicek P,et al. Urinary catecholamines in children with ADHD,modulation by a polyphenolic extract from pine bark(Pycnogenol®). Nutr Neurosci 10: 151–157, 2007

Chapter Fourteen | Top Athletic Performance

Physical exercise dramatically increases oxidation of nutrients to meet the skyrocketing energy demand. Numerous physiological systems and many biochemical interactions are taking place during exercise. Yet, the most significant and performance-limiting interplay is taking place between the cardiorespiratory system and skeletal muscle. This is obvious from the typical 10- to 20-fold increase in inhaled oxygen during exercise over that while resting.

The cardiopulmonary system adjusts to match oxygen and carbon dioxide transport to the metabolic requirements of muscle tissues. (Ref. 1) Increased oxygen demands during physical activity result in a rapid compensatory increase in cardiac output and redistribution of blood flow to skeletal muscles. The blood flow characteristics play a key logistic role for oxygen supply to muscle, return of carbon dioxide to the lungs and delivery of lactic acid to the liver. Only sufficient muscle oxygenation warrants aerobic energy generation and prevents anaerobic build-up of lactic acid. The blood-flow characteristics play a dominant role for peak muscle performance and integrity.

As Pycnogenol® has positive effects on circulation (Refs. 1–4), supplementation with Pycnogenol® can exert a positive influence on physical performance. A better blood supply due to better circulation leads to better oxygenation of the muscles. This improves muscle function so that physical performance increases. This is demonstrated by several controlled studies as discussed later in this chapter.

Top Performance Needs Optimal Blood Flow

Nitric oxide (NO) is the key mediator signaling dilatation of blood vessels to ensure optimal blood flow. NO contributes to increased exercise-induced blood perfusion of organs and plays a key role for coordinating vascular response to exercise. Recent clinical trials have indicated that exercise training is associated with a sustained and systemic increase in the capacity for endothelial nitric oxide production. (Ref. 1) As we discussed in Chapter Two on Heart Disease, Pycnogenol® stimulates the enzyme "endothelial nitric oxide synthase" (eNOS) for enhanced generation of NO from the precursor molecule L-arginine. (Ref. 2)

Pycnogenol® Reduces Exercise-Induced Inflammation and Speeds Healing

In addition to oxygen usage, exercise is also linked to inflammation. Exercise induces an acute inflammation that is needed for muscle repair, overcompensation and muscle growth.

Pycnogenol® was shown to speed up the healing of harmed tissue. (Ref. 3) In this controlled clinical study, 150 mg of Pycnogenol® was taken daily. Thus, Pycnogenol® will significantly support recuperation and may be particularly beneficial for contact sports such as football, wrestling, rugby, or ice hockey.

Pycnogenol® Reduces Cramping and Muscular Pain

Cramping and painful muscles affect essentially every athlete, and may occur during exercise as well as hours later during the recovery phase. Inadequate conditioning and stretching of the muscles is a major cause of muscle cramps. Proper hydration with electrolytes potassium and magnesium is also very important. Cramping and painful muscles are increasingly understood to result from blood circulation reaching its limits in supporting the performing muscle with the necessary oxygen, nutrients, hydration and electrolytes.

Pycnogenol® elevates blood supply to tissues, so a placebo-controlled clinical study with 66 healthy recreational sports people was carried out to investigate benefits for reducing muscle cramping incidence and muscular pain during and after intensive exercise. (Ref. 4)

In this study, the effect of Pycnogenol® was evaluated with subjects daily recording their episodes of cramps and muscular pain at rest and during exercise. The pain level was documented on a visual analogue scale ranging from "total absence of pain" (= 0) to maximum unbearable pain (=10). The baseline values were established during a two-week pretreatment phase. Volunteers took either Pycnogenol® or a placebo for a period of four weeks after which pain level and cramping episodes were estimated again. One week after patients had stopped taking supplementation the effect on muscles was estimated again to find out whether the benefits of Pycnogenol® lasted or a relapse occurred. Subjects were advised to drink at least 1.5 liters of water each day to exclude that insufficient hydration was responsible for any muscle cramps.

Scores for the severity of cramping pain were lowered significantly in both athletes as well as recreational sports people to 13% and 25% of pretreatment values, respectively, following four weeks treatment with Pycnogenol®. After discontinuation of Pycnogenol® intake for a week, a minor and statistically insignificant increase of cramping pain scores was observed.

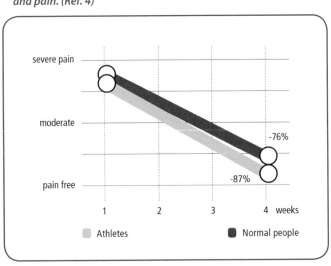

Figure 14.1 Pycnogenol® reduces muscle cramps and pain. (Ref. 4)

The group of healthy recreational subjects experienced a lowered frequency of muscle cramps during exercise and during the recovery phase, which decreased from an average 4.8 incidents per week at baseline to 1.3 after four weeks supplementation with Pycnogenol®.

The group of athletes showed a higher rate of muscle cramps at trial start with an average of 8.6 episodes a week. The frequency of cramp episodes decreased to average 2.4 a week with Pycnogenol®. After discontinuation of Pycnogenol® supplementation for a week the average cramping frequency in all three groups did not increase again suggesting a lasting effect of Pycnogenol®.

The researchers concluded, "Pycnogenol® is effective for reducing pain and cramps in training and re-training, thus increasing the efficiency of training programs both in normal subjects as well as in competitive athletes."

Figure 14.2 Pycnogenol® increases endurance on a treadmill.

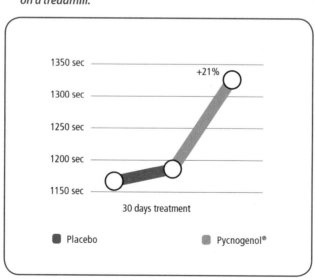

Pycnogenol® Helps Improve Physical Fitness

A study was initiated in Pescara, Italy to determine whether physical fitness increases after supplementation with Pycnogenol®. (Ref. 5) Pycnogenol® as a natural extract is free from anabolic androgenic steroids and stimulants, as has been verified by the Center of Preventive Doping Research at the German Sport University Cologne.

In the first part of the study, 74 healthy, fit subjects received 100 mg Pycnogenol® per day and participated in a training program. During the eight weeks of the study, the subjects completed the US Army physical fitness test (APFT) five times during each week. The APFT consisted of a two-mile run, push ups and sit ups. (Ref. 6)

The control group of 73 subjects, comparable in sex and age distribution, performed the same training program.

Table 14.1 US Army Physical Fitness Results

Test	Females		Males	
	Inclusion	8 weeks	Inclusion	8 weeks
2-Mile Run				
Pycnogenol®	18m	16m	16m	14m
Control	17m	17m	16m	15m
Push-Ups				
Pycnogenol®	31.2	39	56	69
Control	31	34.1	58.2	62.2
Sit-Ups				
Pycnogenol®	61.2	67	63.2	73.2
Control	61.2	64.2	63.2	67.3
Oxidative Stress 1.1.1 Carr units				
Pycnogenol®	339.4	311.3	335.3	318.2
Control	332.3	356.2	310	344.4
*Normal values: males <330 Carr Units; females >335				

At the end of the eight weeks of training, the physical fitness of all participants was significantly improved; however, the Pycnogenol® group achieved significantly better results in each of the tests for both genders.

At the beginning and at the end of the study, all participants were tested for signs of oxidative stress in blood. The oxidative stress after training diminished significantly despite the intensive training in the Pycnogenol® group, but in contrast, participants of the control showed a clear increase of oxidative stress. The data are presented in Table 14.1.

Second study

The second part of the study consisted of a four-week training program for the sprint distance of the triathlon: 750 m swim at open sea, 12 miles cycling and a five-km run.

In the study, 32 male athletes received 150 mg Pycnogenol® daily while 22 male athletes were in the control group without supplementation.

Also in this test, the success of training was observed in all disciplines for the Pycnogenol® group as well as for the control. The results of this study are presented in Table 14.2.

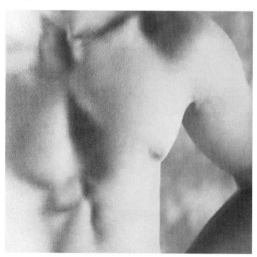

The advantage of Pycnogenol® is evident as can be seen from the Table, especially for the total time to finish the triathlon. Those athletes taking Pycnogenol® supplements decreased their completion time by an average of 10 minutes and 6 minutes faster than placebo group. (Table 14.2)

Table 14.2 Details of triathlon subjects
(32 male athletes, mean age 37.9)

		Inclusion	4 Weeks	Improvement Time (minutes)
Swim	Pycnogenol®	13.28	12.14	
	Control	13.6	12.8	
Bike	Pycnogenol®	39.21	35.07	
	Control	38.9	37.9	
Run	Pycnogenol®	25.04	22.31	
	Control	26.1	24.1	
+ 2 transition times	Pycnogenol®	22.31	20.02	
	Control	22.5	21.5	
TOTAL	Pycnogenol®	100.24	89.44	10.48 min.
	Control	101.1	96.5	4.36 min.
Diff. between groups				6.2 min.
Values in minutes; P=Pycnogenol®; C=Control				

Values for plasma free radicals were obtained before race, at the different transition points of the triathlon and at the end.

Compared to the levels at the beginning of the study, the free radicals in plasma decreased in the Pycnogenol® group after the four weeks, while the values increased in the control group.

Measurements done in blood at the different transition points of the triathlon demonstrated a sharp increase of the free radicals in plasma during the race in the control group, whereas in the Pycnogenol® group, the increase was just moderate. (Table 14.2)

In conclusion, the experience from several clinical trials suggest that Pycnogenol® is effective in protecting the muscle from harm during sports – especially from oxidative stress-enhancing performance, speeding up the recovery, and allowing for faster retraining. The above studies demonstrate that Pycnogenol® is an effective nutritional supplement to maximize sports performance without resorting to the so-called performance-enhancing drugs that are banned in many professional sports.

Chapter References

1. Green DJ, Maiorana A, O'Driscoll G, Taylor R. Effect of exercise training on endothelium-derived nitric oxide function in humans. J Physiol 561: 1–25, 2004

2. Fitzpatrick DF, Bing B, Rohdewald P. Endothelium-dependent vascular effects of Pycnogenol®. J Cardiovasc Pharmacol 32: 509–515, 1998

3. Belcaro G, Cesarone MR, Errichi BM, et al. Venous Ulcers: Microcirculatory Improvement and Faster Healing with Local Use of Pycnogenol®. Angiology 56: 699–705, 2005

4. Vinciguerra G, Belcaro G, Cesarone MR, et al. Cramps and muscular pain: prevention with Pycnogenol® in normal subjects, venous patients, athletes, claudicants and in diabetic microangiopathy. Angiology 57: 331–339, 2006

5. Vinciguerra G, Belcaro G, Bonanni MR, et al. Evaluation of the effects of supplementation with Pycnogenol® on fitness in normal subjects with the Army Physical Fitness Test and in performances of athletes in the 100-minute triathlon. J Sports Med Phys Fitness 53: 644–654, 2013

6. O'Connor JS, Bahrke MS, Tetu RG. 1988 Active Army Physical Fitness Survey. Mil Med 155: 579–585, 1990

Chapter Fifteen | Pycnogenol®
Supplementation

We have discussed many health benefits of this nutritional supplement, Pycnogenol®. Yet, there are many more health applications, such as dental health, that we could include, but we feel that you have already gotten the message. Pycnogenol® improves health! Now it's time to discuss how to best use Pycnogenol® for various health benefits. As a dietary supplement, Pycnogenol® can be used alone or, better, in combination with other dietary supplements. Some supplement combinations are synergistic, providing more health benefits than either nutrient alone.

Combining Pycnogenol® with Other Nutrients

Pycnogenol® is a team player and works hand-in-hand with vitamins. Innumerable combinations containing Pycnogenol® and vitamins are available. However, compared to Pycnogenol®, the vitamins are less potent antioxidants. They are unstable, they are easily oxidized by air. When they are oxidized, they cannot change conformation to catch one free radical more, as Pycnogenol® can do.

Nevertheless, there is no doubt that we need vitamins. So it is very worthwhile that Pycnogenol® offers protection for vitamins and is synergistic with many vitamins.

Pycnogenol® Protects Vitamins C and E

For example, vitamin C, ascorbic acid, becomes quickly converted to dehydroascorbic acid, which can no longer act as a vitamin. Scientists at the University of California, Berkeley found that Pycnogenol®, as a more powerful antioxidant, recycles vitamin C, thus prolonging its usable lifetime. (Ref. 1) In the same way Pycnogenol® protects vitamin E against oxidation in cells. (Ref. 2) Other researchers from Slovakia found that Pycnogenol® acts synergistic in inhibiting the oxidation of plasma proteins together with vitamin C and Trolox, a vitamin E derivative. (Ref. 3)

Synergistic Action with the Vitamin-like Coenzyme Q10

Japanese researchers tested the antioxidant effects of Pycnogenol® and Coenzyme Q10 (CoQ10). (Ref. 4) The protection of lipids against peroxidation was unexpectedly high, when both agents acted in concert, compared to the results obtained with Pycnogenol® and CoQ10 alone. The combination of the water-soluble Pycnogenol® with the lipophilic CoQ10 allowed the mixture to produce a better protection of the sensitive cell membranes.

In a clinical study, patients with moderate cardiac defect, were supplemented with a combination of Pycnogenol® with CoQ10 or placebo. The patients continued to take their prescribed medication in addition to the supplementation. (Ref. 5)

Relative to the control group who received a placebo along with their medication, patients receiving both 105 mg Pycnogenol® plus 350 mg CoQ10 could walk a 3.3 times longer distance on a treadmill, their quality of life index changed from "patients are handicapped and dependent on qualified medical help" towards "help and medical assistance are often required." No such improvement was seen in the controls.

The ejection fraction of the heart, a measure of the function of the heart as a pump, increased significantly by 22% with Pycnogenol® plus CoQ10. Heart rate and blood pressure were lowered. (Ref. 5) A comparison with the effects of Pycnogenol® and CoQ10 as single components could not be done, so that we don't know whether these good results are based on a true synergistic effect of Pycnogenol® and CoQ10.

Synergism with Lutein, a Substance Related to Vitamin A

The lipids in cells of the retina are very sensitive to oxidation. They are protected inside the eye by a number of antioxidants. Japanese researchers exposed cells from the retina to oxidative stress, leading to peroxidation of the lipids of the retinal cells. (Ref. 6) The addition of lutein had little effect, whereas Pycnogenol® inhibited peroxidation significantly. Remarkably, the combination of lutein with Pycnogenol® produced a much higher effect as could be expected from the contributions of the single components. As in the foregoing example with Pycnogenol® and CoQ10, it is the hand-in-hand action of a lipophilic (lutein) and a hydrophilic (Pycnogenol®) component that greatly enhances the antioxidant effect.

Optimal Dosages of Pycnogenol®

As with any nutrient, different people need different amounts of Pycnogenol® for best health. Healthy teenagers need less than the seriously ill. The following may help you decide how much Pycnogenol® is best for you.

The following dosages of Pycnogenol® have been used and studied in scientific research:

In clinical studies, daily amounts from 50 mg up to 300 mg Pycnogenol® have been used. Generally speaking, acute problems such as hemorrhoids or exposure to allergens require a dose between 200 and 300 mg. For prevention and regulation of mild symptoms, doses of two mg per lb (1mg/kg) body weight are recommended.

Clinical results from Japanese studies showed that 60 mg Pycnogenol® per day where already effective in Japanese people, most probably due to their lower body weight.

If you suffer from severe symptoms such as menopausal pain, joint pain, depression, asthma, or ADHD, you can use a higher dosage of Pycnoge-

Condition	Dosage Pycnogenol / Day	Dosage Regimen
Allergies / Hay fever	100 mg	50 mg twice daily
Asthma in children	1 mg/pound of body weight or 2mg/kg bodyweight	given in two divided doses
Poor circulation	50 - 360 mg	50 - 100 mg 1-3 times daily
Retina diseases, including those related to diabetes	150mg	50 mg three times daily
Mild high blood pressure	200 mg	100 mg twice daily
Improving exercise capacity in athletes	200 mg	100 mg twice daily

nol® in the beginning. When the symptoms get better you might reduce the dosage of Pycnogenol®. So you might start with a supplementation of 200 mg of Pycnogenol® per day to get symptoms under control as soon as possible. After the symptoms decline, you might try decreasing the dosage.

Safety of Pycnogenol®

Pycnogenol® has been extensively researched for its safety. In the USA, Pycnogenol® has GRAS (generally recognized as safe) status. By the way, a onetime overdosing of Pycnogenol® is not a big issue and is not dangerous. Single dosages from 3 grams up to 5.4 grams have been tolerated without any problem. However, permanent intake of 300 mg Pycnogenol® and more may lower blood pressure and / or blood sugar levels below normal values and requires control of blood glucose and blood pressure.

In double-blind and open clinical studies, about 7,000 patients and healthy subjects have reported no unwanted effects. In the majority of the studies, participants did not complain about any side effects of Pycnogenol®. Seldom, mild unwanted symptoms such as headache, gastric discomfort, dizziness, and nausea were noted. People who may experience gastric discomfort are advised to take Pycnogenol® with or after a meal.

No serious side effects had been reported so far, despite the fact that millions of doses Pycnogenol® have been sold worldwide over the last 40 years.

So we can conclude that Pycnogenol® is a very safe supplement with a long history of usage and safety.

How about long-term intake of Pycnogenol®? In clinical studies that lasted up to six months of Pycnogenol® supplementation, no side effects occurred.

Conclusion

Pycnogenol® is not a pharmaceutical. Pycnogenol® is a safe supplement that has beneficial effects in certain health conditions. However, the effects are not as strong and fast as the effects of pharmaceutical drugs. Pycnogenol® helps in a natural and safe way to control symptoms of some diseases, but in severe cases, it cannot replace drugs. So, for example, it cannot block an acute asthma attack, it is not able to lower really high blood pressure, and it has not the power to bring immediate pain relief in an acute inflammation.

One role of Pycnogenol® is to help you to get back into a healthy status from mild or moderate discomfort. However, the foremost job of Pycnogenol® is to hold body and soul in balance. The daily intake of Pycnogenol® is the road to keep and hold you in good shape: Look, feel and live better.

We hope this book gives you the information you need to make science-based decisions about Pycnogenol® and your health. We have shared our experience with Pycnogenol® with you and have given you the data and studies to enable you to make your independent decisions.

Good Health to You and May You Live Better Longer!

Chapter References

1.	Cossins E, Lee R, Packer L. ESR studies of vitamin C regeneration, order of reactivity of natural source phytochemical preparations. Biochem Mol Biol Int 45: 583–597, 1998

2.	Virgili F, Kim D, Packer L. Procyanidins extracted from pine bark protect α-tocopherol in ECV 304 endothelial cells challenged by activated RAW 264.7 macrophages: role of nitric oxide peroxynitrite. FEBS Lett 431: 315–318, 1998

3.	Sivonova M, Zitnanova I, Horakova L, et al. The combined effect of Pycnogenol® with ascorbic acid and Trolox on the oxidation of lipids and proteins. Gen Physiol Biophys 25: 379–396, 2006

4.	Chida M, Suzuki K, Nakanishi-Ueda T, et al. In vitro testing of antioxidants and biochemical end-points in bovine retinal tissue. Ophtalmic Res 31: 407–415, 1999

5.	Belcaro G, Cesarone MR, Dugall M, et al. Investigation of Pycnogenol® in combination with coenzyme Q10 in heart failure patients (NYHA II/III). Panminerva Med 52: 21–25, 2010

6.	Nakanishi-Ueda T, Kamegawa M, Ishigaki S, et al. Inhibitory effect of Lutein and Pycnogenol® on lipid peroxidation in porcine retinal homogenate. J Clin Biochem Nutr 38: 204–210, 2006

INDEX